lditive Innovations - Evolving Consumer Expectations and Industrial Needs
dx.doi.org/10.5772/intechopen.1009736
by Fernanda Cosme, Fernando M. Nunes and Miguel Ribeiro

butors
Kallampilly, Aswathy Thankappan, Din Mohammad Mushfiq, Fernanda Cosme, Fernando M. Nunes,
am Manoharan, Juliana Milheiro, Luís Filipe-Ribeiro, Luís Moreira, Miguel Ribeiro, Mário
'a, Nawab Nasir, Ramabhau Tumadu Patil, Ravindra Naik, Sandrine Ferreira, Sevinç Ateş, Shahnaz
Sweety, Tahsin Ahmed Rupok, Victoria Mukuni

irst published in London, United Kingdom, 2026 by IntechOpen
ntechOpen is the global imprint of INTECHOPEN LIMITED, registered in England and Wales,
egistration number: 11086078, 167-169 Great Portland Street, London, W1W 5PF, United Kingdom

For EU product safety concerns: IN TECH d.o.o., Prolaz Marije Krucifikse Kozulić 3, 51000 Rijeka,
Croatia, info@intechopen.com or visit our website at intechopen.com.

British Library Cataloguing-in-Publication Data
A catalogue record for this book is available from the British Library

Food Additive Innovations - Evolving Consumer Expectations and Industrial Needs
Edited by Fernanda Cosme, Fernando M. Nunes and Miguel Ribeiro
p. cm.

This title is part of the Food Science and Nutrition Book Series, Volume 25
Topic: Food Technology
Series Editor: Maria Rosário Bronze
Topic Editor: Carlos Dias Pereira

Print ISBN 978-1-83635-914-2
Online ISBN 978-1-83635-913-5
eBook (PDF) ISBN 978-1-83635-915-9
ISSN 2977-8174

If disposing of this product, please recycle the paper responsibly.

Food Additive Innovation - Evolving Consumer Expectations and Industrial Needs

Edited by Fernanda Cosme, Fernando M. Nunes and Miguel Ribeiro

Published in London, United Kingdom

IntechOpen Book Series

Food Science and Nutrition

Volume 25

Aims and Scope of the Series

The significance of food is undeniable, especially in light of the impending challenge facing humanity: ensuring there will be enough food to meet the basic needs of a population expected to reach approximately 10 billion by 2050. These food-related challenges align with some of the United Nations' sustainable development goals, with a target to achieve them by 2030. One thing is certain: food should be not only nourishing and safe but also tailored to the diverse needs of individuals throughout their lifetimes, all while meeting consumers' sensory expectations. Understanding the diverse chemical composition of food, often referred to as biodiversity, and how these components can contribute to human health by considering factors like bioaccessibility, bioavailability, and bioactivity at the organ level, is crucial for grasping and promoting a healthy diet. Thanks to the continuous evolution of analytical methods and interdisciplinary research, significant strides have been made in the field of food science and nutrition.

Meet the Series Editor

Maria Rosário Bronze has been working in Analytical Chemistry since 1986. Her Ph.D. in 1999 contributed to the study of food products using capillary electrophoresis. The main goal of her research since 1999 has been focused on Analytical Chemistry applied mainly to the analysis of foods and by-products of food industry. She conducted research in collaboration with national and international research groups, at iBET and ITQB Technology Division. From 2017 until 2021 she was head of Food & Health Division at iBET and head of the Food Functionality and Bioactives Laboratory. MR Bronze has been an Associate Professor at the Pharmacy Faculty of Lisbon University and head of the Structural Analysis Laboratory since 2012. As a researcher, MR Bronze is a Senior Scientific Advisor at Food & Health Division at iBET and Head of Food Functionality and Bioactives Laboratory at the same Institute, Collaborator at iMED and Researcher at ITQB NOVA. Her current research is focused on quality and beneficial health effects of food components. Gas and liquid chromatography associated with mass spectrometry are used by MR Bronze in the characterization of samples. Sensory evaluation is also an important area of her research. The main food products studied by her are olive tree products (olive, olive oil, leaves), cereals such as maize, legumes (faba bean, pea, chickpea, lentils) fruits (apple, grapes, opuntia ficus), fruit juices and wine, among others. More recently her interests have also involved biodiversity, bioaccessibility, and bioavailability studies on food products and their components, mainly phytochemicals as phenolic compounds, using different analytical tools such as mass spectrometry. As a senior scientific advisor at Food & Health Division at iBET she is involved in different areas: (i) isolation, characterization and formulation of bioactive and functional compounds or extracts from natural sources and wastes from food and other related industries; (ii) pre-clinical assays to provide support to understand health claims related with the beneficial effects of food nutrients/bioactive components; (iii) establishment of analytical methodologies including mass spectrometry state-of-the-art to fully characterize different matrices, from food products, natural extracts or biological fluids (Food Functionality and Bioactives Laboratory).

Meet the Volume Editors

Fernanda Cosme is an Associate Professor at the University of Trás-os-Montes and Alto Douro, with research focused on grapes and wines, including grape composition, wine technology, wine stabilization, quality and safety, and wine aging. She has served as scientific editor for four books in enology and is the author of patents and numerous research articles published in international journals indexed in the Science Citation Index in the area of food science. She has also published book chapters and technical publications and participates in international and national congresses as a speaker. She has participated in funded research projects, including projects with industrial partners, and supervised Ph.D. and MSc students in food science and enology (orcid. org/0000-0001-5924-1050; Scopus ID: 24437875500).

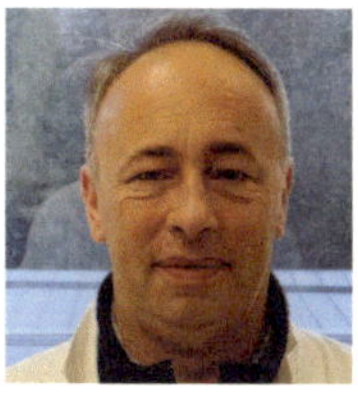

Fernando M. Nunes is a Full Professor in Analytical and Food Chemistry at the University of Trás-os-Montes and Alto Douro. He has experience in the analysis of phenolic compounds and wine aromas, carbohydrate analysis, Maillard reactions, melanoidins, and food additives. He is the author of 18 patents and 184 articles in international journals, all cited in the Science Citation Index, in the areas of food chemistry, carbohydrate chemistry, polysaccharide chemistry, phenolic compounds, and the biological activities of these compounds. He has been the principal investigator of several research projects with industrial partners (orcid. org/0000-0001-5540-318X; Scopus ID 7102392847).

Miguel Ribeiro is a researcher in food biotechnology, food safety, and innovation, with expertise in molecular biology, proteomics, and biotechnology. He obtained his Ph.D. in Technological, Comparative, and Molecular Genetics from the University of Trás-os-Montes and Alto Douro in 2016. His research focuses on biobased solutions to improve food quality, safety, and sustainability, particularly through gluten-detoxification technologies that enhance the quality of life for individuals with celiac disease. He has contributed to studies on gluten-specific nanoparticles, acrylamide reduction in foods, bioactive compounds from mushrooms, and sustainable protein stabilization in wine production. He demonstrates a strong commitment to scientific excellence, innovation, and technology transfer, with more than 40 indexed publications, over 40 conference presentations, and 6 patent applications.

Contents

Preface

Food additives have played a fundamental role in the development of food systems, enabling the production of safe, stable, nutritious, and appealing foods for a growing global population. For decades, additives have contributed to food preservation, quality enhancement, and product innovation, supporting both consumer needs and industrial efficiency. However, consumer expectations have evolved significantly in recent years. Today, consumers seek foods that not only meet nutritional requirements but also align with broader concerns related to health, transparency, sustainability, and environmental responsibility. As a result, the science and application of food additives are undergoing a profound transformation.

This book explores the field of food additives, highlighting innovations, technological advances, and regulatory considerations shaping the future of the food industry. It brings together current knowledge on both traditional and emerging additives, providing readers with a comprehensive understanding of their functions, benefits, limitations, and applications.

A central theme throughout the book is the growing shift toward clean-label and naturally derived ingredients. Increasing consumer demand for minimally processed foods has encouraged the development of alternatives to conventional additives, including plant-based antioxidants, natural sweeteners, and functional ingredients such as probiotics, prebiotics, and dietary fibers. These innovations not only help maintain food quality and extend shelf life but also contribute to improved health and well-being, reflecting the expanding role of food additives beyond preservation and sensory enhancement.

Emerging developments in next-generation food additives are examined, highlighting innovative compounds designed to enhance functionality and consumer acceptance. The growing importance of natural sweeteners as alternatives to conventional sugar and artificial sweeteners is also addressed, together with the role of plant-derived antioxidants in improving food stability, extending shelf life, and contributing to health-promoting properties.

On the other hand, specialized applications are discussed, including additive-based wine tartaric stabilization strategies used in enology to maintain product quality and consumer acceptance.

Another important focus of this book is the integration of food additives within innovative packaging systems. Smart packaging technologies represent a promising frontier in food preservation, combining active and intelligent components that enhance food safety, extend shelf life, improve traceability, and provide real-time information about product quality.

The future of food additives lies not only in technological advancement but also in ethical responsibility. The next generation of additives will likely be more personalized, sustainable, and derived from renewable or upcycled resources. Achieving these goals will require collaboration among scientists, food technologists, regulatory agencies, industry leaders, and consumers. Through such cooperation, it will be possible to develop innovative solutions that enhance food quality, promote health, reduce environmental impact, and contribute to more sustainable food systems.

We hope that this book will serve as a valuable resource for understanding the innovation and responsibility that define food additives and inspire further research and development in this important field.

Fernanda Cosme, Fernando M. Nunes and Miguel-Ribeiro
Chemistry Research Centre - Vila Real, (CQ-VR),
Food and Wine Chemistry Laboratory (FoodWin),
University of Trás-os-Montes and Alto Douro,
Vila Real, Portugal

Chapter 1

Next-Generation Food Additives: Innovations for Health, Clean Labels and Functional Benefits

Shahnaz Parvin Sweety and Tahsin Ahmed Rupok

Abstract

This chapter explores the recent innovations in food additives that align with changing consumer expectations and pressing industrial demands. With growing awareness of health, sustainability, and transparency, there is a shift toward clean-label additives – simple, natural ingredients that maintain food quality without excessive processing. The chapter delves into plant-based antioxidants, natural emulsifiers, and fermentation-derived biopreservatives, highlighting their role in meeting the clean-label movement. It also examines low-calorie sugar alternatives, potassium-based salt substitutes, and functional additives such as probiotics, prebiotics, and dietary fibers aimed at enhancing well-being. Additionally, the impact of nanotechnology, particularly nanoencapsulation, is discussed in the context of controlled nutrient release and improved stability. Finally, the chapter emphasizes the importance of safety and regulatory compliance, showcasing how innovation must align with strict standards. This chapter provides a foundation for understanding the future of food additive development in a rapidly evolving landscape.

Keywords: Clean-Label additives, food innovation, biopreservatives, sugar substitutes, functional ingredients, nanotechnology, probiotics, food safety, regulatory compliance

1. Introduction

Food additives have long played a pivotal role in the food industry, serving functions that range from preservation and texture modification to flavor enhancement and nutrient fortification. Historically, synthetic additives dominated food processing due to their efficacy, low cost, and scalability. However, a significant shift is now underway, driven by evolving consumer expectations, regulatory reforms, and scientific advancements. Consumers increasingly demand transparency, health-conscious choices, and sustainable practices – trends that have given rise to the clean-label movement and fueled innovation in food additive development [1, 2].

The concept of a “clean label” encompasses ingredients that are natural, minimally processed, easily recognizable, and free from synthetic chemicals. This consumer-driven phenomenon has redefined how food companies formulate products, prompting a reevaluation of additive sources, functions, and long-term health effects [3]. In fact, market research reports indicate that over 70% of global consumers consider clean-label claims when making purchasing decisions, with natural preservatives, colors, and sweeteners being key focus areas [4, 5].

At the heart of this transformation are next-generation additives that provide health benefits while maintaining or enhancing food quality. Plant-derived antioxidants, for example, offer protection against oxidative spoilage while contributing to the nutritional value of the product. Extracts from rosemary, grape seed, green tea, and pomegranate are being extensively researched and applied due to their natural antioxidant and antimicrobial properties [6, 7]. Similarly, natural emulsifiers derived from plant proteins, saponins, and polysaccharides are replacing traditional synthetic emulsifiers, offering better compatibility with clean-label formulations [8, 9].

Biotechnological advancements have enabled the use of fermentation-derived biopreservatives such as nisin and natamycin, which inhibit spoilage microorganisms in dairy, meat, and beverage applications [10]. At the same time, public health campaigns promoting sugar and sodium reduction have accelerated the adoption of natural sweeteners like stevia, monk fruit, and rare sugars such as allulose, along with potassium-based salt substitutes [11, 12]. These alternatives allow food manufacturers to meet nutritional guidelines without compromising taste or shelf stability.

Functionality has also emerged as a key area of innovation. Probiotic and prebiotic additives are being integrated into diverse food matrices to support gut health, while dietary fibers are utilized not only for health benefits but also for improving texture and satiety [13, 14]. In parallel, nanotechnology has enabled controlled delivery of these functional ingredients through nanoencapsulation, which enhances stability, bioavailability, and sensory profiles [15, 16].

This chapter explores these developments in depth, highlighting the scientific rationale, technological applications, and regulatory considerations behind each innovation. With a focus on health, sustainability, and functionality, it provides for understanding of how food additives are being reimagined to meet the needs of a rapidly changing global food system. According to **Figure 1.** we saw the flowchart of Next-Generation Food Additives Innovation Strategy.

2. The clean label movement: Shifting consumer priorities

2.1 What is clean label?

The term clean label refers to food products made with minimal, natural, and easily recognizable ingredients – free from artificial colors, flavors, preservatives, and synthetic chemicals. Consumers are no longer passive; they now demand transparency in ingredient sourcing, processing methods, and health implications. This shift reflects broader societal trends toward wellness, environmental responsibility, and digital empowerment.

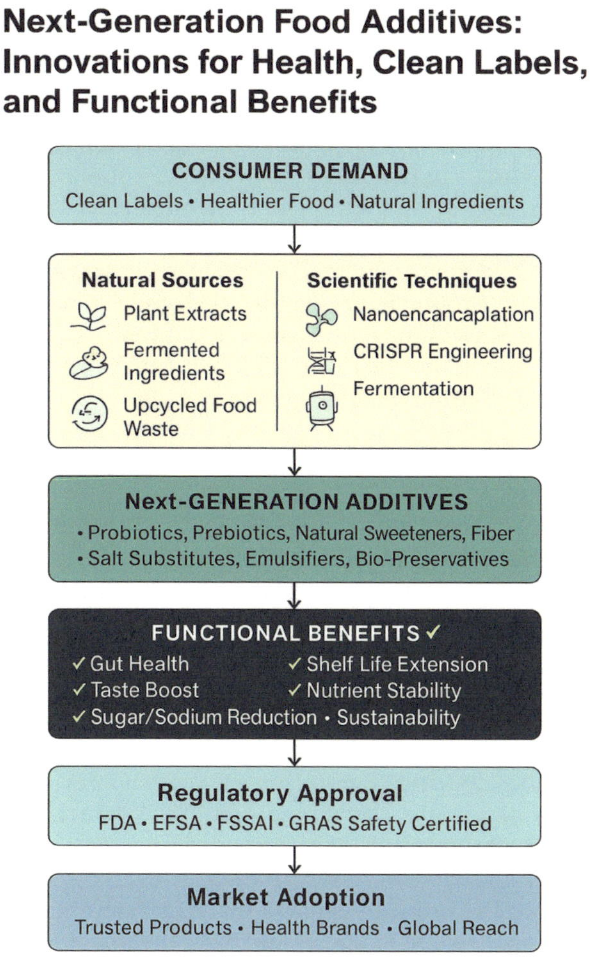

Figure 1.
Strategy of Next-Generation Food Additives Innovation.

According to a 2022 Global Data survey, 73% of consumers stated they are more likely to purchase products with fewer synthetic additives and simpler ingredient lists. Clean label is no longer a niche trend but a mainstream market force reshaping product development strategy across the food industry [17].

2.2 Drivers of the clean-label trend

The clean-label movement is being driven by a combination of health, ethical, and technological factors that are reshaping consumer expectations. Increasing health awareness, particularly due to the growing prevalence of non-communicable diseases such as obesity, type 2 diabetes, and cardiovascular disorders, has made consumers more cautious about what they eat. As a result, there is a strong preference for foods made with simpler, more natural ingredients rather than synthetic additives [18]. This trend is further reinforced by the rising incidence of food allergies and sensitivities, prompting individuals to avoid artificial substances like monosodium glutamate (MSG), synthetic dyes, and sulfites that may trigger adverse reactions. Transparent and trustworthy labeling has thus become essential in helping consumers make safer food choices and reduce anxiety around hidden additives [18, 19]. Additionally, the digital era has introduced new tools for transparency – mobile applications such as Yuka and Open Food Facts allow users to instantly assess a product's nutritional value, additive content, and organic status. This real-time access to information pressures manufacturers to reformulate products with cleaner, more consumer-friendly ingredients. Beyond health and transparency, ethical and environmental consciousness plays a significant role [20]. Younger generations, in particular, are aligning their food choices with values related to sustainability, animal welfare, and responsible sourcing. Consequently, plant-based and naturally derived products are gaining popularity as symbols of both personal wellness and environmental responsibility [18, 21].

2.3 Industrial implications

The clean-label movement is reshaping the food industry by driving manufacturers to reformulate products that balance functionality, safety, and consumer expectations. This transformation presents both challenges and opportunities across the food supply chain. One of the major challenges lies in reformulation itself. Replacing synthetic additives with natural alternatives often compromises shelf life, microbial stability, and sensory consistency. For example, removing artificial preservatives can accelerate spoilage, while eliminating synthetic colorants may reduce visual appeal. Achieving the same level of quality with natural substitutes demands significant research and development investment, particularly to maintain texture, flavor, and cost-effectiveness without the support of conventional synthetic agents [22].

At the same time, this trend is fostering innovation across the food and biotechnology sectors. The growing demand for natural alternatives has accelerated the development of plant-based extracts, microbial metabolites, and fermentation-derived compounds that can effectively mimic the roles of traditional additives. Notable examples include natural antimicrobials such as nisin and natamycin, colorants like anthocyanins and carotenoids, and enzymes or dietary fibers that serve as stabilizers and texturizers. Recent advances in biotechnology, nanotechnology, and green extraction technologies are further enhancing the scalability and efficiency of producing these natural compounds [23].

However, companies must also navigate complex regulatory and labeling frameworks to ensure transparency and compliance. Regulatory agencies such as the FDA (Food and Drug Administration), EFSA (European Food Safety Authority), FSSAI (Food Safety and Standards Authority of India), and JECFA (Joint FAO/WHO Expert Committee on Food Additives) define and monitor claims like "natural," "organic," and "additive-free" differently across regions, compelling manufacturers to balance marketing strategies with legal and ethical obligations [24]. Despite these hurdles, the clean-label trend presents substantial market opportunities. Consumers' willingness to pay a premium for products perceived as healthier and more transparent has made clean-label foods a lucrative market segment. Moreover, reformulating products in line with clean-label principles not only meets consumer expectations but also strengthens brand differentiation and trust – factors that are particularly valuable in competitive categories such as beverages, dairy, and bakery products [1].

The clean-label movement is reshaping the food industry by pushing manufacturers to reformulate products while balancing functionality, safety, and consumer expectations. This trend presents both challenges and opportunities across the food supply chain.

3. Innovations in natural additives

3.1 Plant-based antioxidants

Antioxidants are used to prevent oxidation, thus extending shelf life and preserving flavor, especially in fat-rich foods. **Table 1** highlights four widely used natural antioxidants derived from plant sources, each containing potent bioactive compounds, including carnosic acid, catechins, proanthocyanidins, and curcumin. As shown in **Table 1**, these extracts are applied across diverse food categories, including oils, beverages, snacks, and health products, to enhance stability, extend shelf life, and provide added functional benefits. Natural antioxidants not only protect food but are also linked with health benefits, including reduced inflammation and oxidative stress [25–28].

3.2 Natural emulsifiers

Emulsifiers allow water and oil to blend, maintaining texture and stability in foods like mayonnaise, ice cream, and plant-based milks. **Table 2** presents four emerging clean-label emulsifiers sourced from plants, including phospholipids, polysaccharides, proteins, and saponins. As shown in **Table 2**, each emulsifier offers specific functional advantages – such as viscosity, gel formation, allergen-free formulation, and foaming capacity – making them suitable for a wide range of food

Natural antioxidant	Bioactive compound	Applications	Ref.
Rosemary extract	Carnosic acid, rosmarinic acid	Oils, meats, snacks	[29]
Green tea extract	Catechins (Epigallocatechin gallate, EGCG)	Beverages, bakery, dairy	[26]
Grape seed extract	Proanthocyanidins	Chocolates, dressings, cereals	[30]
Turmeric	Curcumin	Sauces, health drinks	[31]

Table 1.
The sources, bioactive compounds, and applications of natural antioxidants.

Emulsifier	Source	Type	Advantages	Ref.
Lecithin	Soy, sunflower	Phospholipid	Vegan, clean-label	[32]
Chia seed mucilage	Chia seeds	Polysaccharide	High viscosity, gel-forming	[33]
Pea protein isolate	Peas	Protein	Allergen-free alternative to soy	[34]
Quillaja saponins	Soapbark tree	Saponin	Foaming properties in beverages	[35]

Table 2.
The sources, types, and advantages of natural emulsifiers.

applications. These emulsifiers are biodegradable, functional, and compatible with vegan and allergen-free diets.

3.3 Fermentation-derived preservatives

Microbial fermentation can yield potent, safe preservatives. **Table 3** summarizes three microbial-based preservatives, each derived from beneficial microorganisms and used to enhance food safety. As shown in **Table 3**, these compounds work through different mechanisms, such as pore formation, antifungal action, or acidity-driven antimicrobial effects, making them effective across diverse products like cheese, fermented vegetables, and bakery items. These additives align with clean-label values and are considered GRAS (Generally Recognized As Safe) by the FDA (United States Food and Drug Administration) [36–38].

4. Functional additives: Probiotics, prebiotics, and dietary fibers

Functional additives go beyond preservation – they enhance overall health, particularly gut health, metabolic function, and immune support.

4.1 Probiotics

Probiotics are live microorganisms that, when administered in adequate amounts, confer health benefits on the host [41]. Common probiotic strains include *Lactobacillus rhamnosus*, *Lactobacillus acidophilus*, and *Bifidobacterium bifidum*, each known for their ability to support digestive and immune health. These beneficial microbes are widely incorporated into various food products such as yogurt, kefir, functional beverages, and infant formulas. Beyond their nutritional role, probiotics help maintain a balanced gut microbiota, reduce intestinal inflammation, and prevent antibiotic-associated diarrhea. Their incorporation into everyday foods not only enhances the functional value of such products but also contributes to overall health promotion and disease prevention.

4.2 Prebiotics

Prebiotics are non-digestible food components, primarily dietary fibers, that selectively promote the growth and activity of beneficial bacteria in the gut. Common

Preservative	Microbial origin	Mechanism	Use case	Ref.
Nisin	*Lactococcus lactis*	Pore formation in bacterial membranes	Cheese, sauces, canned soups	[39]
Natamycin	*Streptomyces natalensis*	Antifungal	Dairy, bakery	[40]
Lactic acid	*Lactobacillus*	pH reduction, antimicrobial	Pickles, fermented vegetables	[36]

Table 3.
The microbial origin, mechanism, and uses of natural preservatives.

sources of prebiotics include inulin derived from chicory root, as well as fructooligosaccharides (FOS) and galactooligosaccharides (GOS), which are naturally present in various plant-based foods. These compounds are frequently incorporated into high-fiber cereals, energy bars, and functional beverages to enhance their nutritional and digestive value. The consumption of prebiotics supports improved digestion, enhances calcium absorption, and helps reduce gut inflammation, thereby contributing to overall gastrointestinal health and well-being [13].

5. Low-Calorie sugar alternatives and salt substitutes

Consumers seek sweet and savory flavors without compromising health. Innovations in sugar and salt reduction are reshaping processed food development.

5.1 Natural sweeteners

These natural sweeteners offer alternatives to traditional sugar, catering to various dietary needs and preferences. **Table 4** provides an overview of three popular low-calorie and natural sweeteners, comparing their sweetness intensity, energy value, and functional properties. As shown in **Table 4**, Stevia and Monk Fruit offer exceptionally high sweetness with zero calories, while Allulose provides a sugar-like taste with very low caloric content and minimal impact on blood glucose.

5.2 Salt substitutes

Excess sodium is linked with hypertension and cardiovascular diseases. **Table 5** highlights three clean-label salt substitutes, each offering a unique way to reduce sodium while maintaining flavor. As shown in **Table 5**, potassium chloride

Sweetener	Relative sweetness (vs. sucrose)	Caloric value (kcal/g)	Properties	Ref.
Stevia	~200–300 ×	0	Zero glycemic index; suitable for diabetics; may aid in blood glucose control.	[42]
Monk fruit	~100–250 ×	0	Shows antioxidant activity; does not affect blood sugar levels.	[43]
Allulose	~70%	~0.2–0.4	Sugar-like taste; minimal effect on blood glucose.	[44]

Table 4.
Some sugar alternatives along with their properties.

Substitute	Composition	Key advantage	Ref.
Potassium chloride	KCl	Lowers blood pressure, balances Na	[45]
Seaweed extract	Natural umami enhancers	Adds flavor, minerals	[46]
Yeast extract blends	High in glutamates	Savory taste, 30–40% salt reduction	[47]

Table 5.
The composition and key advantages of salt substitutes.

provides mineral balance, seaweed extract contributes natural umami and trace nutrients, and yeast extract blends deliver savory notes that enable up to 40% salt reduction.

6. Nanotechnology in food additives: Smart delivery and enhanced stability

6.1 Introduction to nanoadditives

Nanotechnology involves manipulating materials at the nanoscale (1–100 nm) to enhance physical, chemical, and biological functions. In the food industry, nanotechnology is a game changer in additive formulation, offering improved solubility, protection of sensitive compounds, controlled release, and better bioavailability [48]. Nanoencapsulation is the most widely applied method – encasing functional ingredients like vitamins, flavors, or antioxidants inside biodegradable carriers for targeted delivery and prolonged shelf life.

6.2 Applications of nanoencapsulation

Table 6 presents four nano-enabled delivery systems that improve the stability, absorption, and functionality of key bioactive ingredients in foods. As shown in **Table 6**, different nano carriers – such as liposomes, nano-emulsions, solid lipid nanoparticles, and alginate nanogels – help overcome challenges like oxidation, poor solubility, low uptake, and probiotic viability.

6.3 Safety and regulatory considerations

Despite the great potential of nano-additives in enhancing food quality and functionality, safety remains a major concern. The use of nanomaterials in food raises questions about possible risks such as unknown toxicity, bioaccumulation in tissues, and unforeseen interactions with biological cells that could affect human health [53, 54]. Recognizing these concerns, regulatory agencies around the world have established specific guidelines to ensure consumer safety. In Europe, the European Food Safety Authority (EFSA) requires a comprehensive toxicological

Active ingredient	Nano carrier	Functionality	Application	Ref.
Vitamin D	Liposomes, chitosan	Protects from oxidation, enhances uptake	Dairy, beverages	[49]
Omega-3 fatty acids	Nano-emulsions, cyclodextrin	Masks fishy odor, improves solubility	Infant formula, spreads	[50]
Curcumin	Solid lipid nanoparticles	Enhances stability and absorption	Supplements, health drinks	[51]
Probiotics	Alginate nanogels	Increases viability in gut environment	Yogurt, functional bars	[52]

Table 6.
The nano carrier system, functionality, and applications of different active ingredients.

evaluation and clear labeling of any engineered nanomaterials used in food products [55]. Similarly, in the United States, the Food and Drug Administration (FDA) has issued detailed guidance for nano-formulated food ingredients, emphasizing Generally Recognized as Safe (GRAS) evaluations and premarket reviews for new or modified uses [56].

7. Safety, compliance, and global regulations

Ensuring the safety of food additives is central to modern food policy, as these substances directly affect public health, consumer trust, and international trade. Food additives, whether synthetic or naturally derived, undergo stringent pre-market evaluations that assess their toxicological profile, metabolic fate, dietary exposure, and potential long-term effects [57, 58]. International regulatory authorities such as the FDA, EFSA, FSSAI, and JECFA have developed scientific risk-assessment frameworks to ensure only safe levels of additives are permitted in the food supply.

7.1 Why food additive safety matters

Food additives are essential for maintaining the quality, safety, and shelf life of modern food products; however, their improper use can lead to significant health concerns. Instances such as the restriction of certain synthetic colorants linked to hyperactivity in children and the European Union's ban on titanium dioxide highlight the need for continuous regulatory vigilance. Ensuring food additive safety is crucial for three main reasons. First, health protection relies on rigorous safety testing to confirm that additives do not cause toxic, allergenic, carcinogenic, or metabolic effects when used within approved limits [57, 59]. Second, consumer confidence depends on transparent safety assessments and regulatory oversight, which have become increasingly vital in response to growing clean-label and natural product trends [57]. Finally, global trade benefits from harmonized safety standards that enable the international exchange of approved additives under Codex Alimentarius guidelines, promoting consistency and consumer protection across borders[60].

7.2 Regulatory authorities and guidelines

Table 7 outlines major global regulatory bodies responsible for ensuring the safety, assessment, and standardization of food additives across different regions. As shown in **Table** 7, organizations such as the FDA, EFSA, FSSAI, and JECFA play crucial roles in setting safety limits, issuing risk assessments, and establishing internationally recognized regulatory frameworks.

7.3 Risk assessment parameters

Risk assessment provides a systematic framework for determining the safe levels of food additive consumption. One of the key parameters is the Acceptable Daily Intake (ADI), which defines the maximum amount of a substance that can be safely consumed every day over a lifetime without posing health risks [65]. Another

Regulatory body	Region	Key role	Ref.
FDA (Food and Drug Administration)	USA	Oversees GRAS designation, labeling, safety reviews	[61]
EFSA (European Food Safety Authority)	EU	Publishes risk assessments and ADI levels	[62]
FSSAI (Food Safety and Standards Authority of India)	India	Establishes permissible limits for food additives	[63]
JECFA (Joint FAO/WHO Expert Committee on Food Additives)	Global	Develops international food additive standards	[64]

Table 7.
The region and key role of different regulatory body.

important measure is the No-Observed-Adverse-Effect Level (NOAEL), representing the highest tested dose in animal studies that shows no detectable harmful effects [66]. Additionally, the Threshold of Toxicological Concern (TTC) offers a risk-based approach for evaluating substances that have very low exposure levels or limited toxicological data, ensuring safety even when comprehensive studies are unavailable [67]. A notable example of how these parameters guide regulatory action is the re-evaluation of titanium dioxide (E171). Between 2021 and 2023, the European Food Safety Authority (EFSA) reassessed this commonly used whitening agent and, due to emerging evidence of potential genotoxicity, concluded that it could no longer be considered safe for food use. Consequently, titanium dioxide was banned across the European Union. This case highlights that risk assessment is not static but rather an evolving process that must adapt to new scientific findings and updated safety data [68].

8. Next-generation food additives: Clean label and functional innovation

The exploration of next-generation food additives signifies a major transformation in the global food industry – one that is increasingly driven by consumer priorities such as health, transparency, and sustainability rather than by cost or convenience alone. This shift reflects a broader reorientation toward clean, functional, and ethically sourced ingredients that align with the values of modern consumers.

A key aspect of this transformation lies in the comparison between natural and synthetic additives. Conventional synthetic additives such as Butylated Hydroxyanisole (BHA), Butylated Hydroxytoluene (BHT), Monosodium Glutamate (MSG), and artificial sweeteners like aspartame have long been favored for their cost-effectiveness, consistency, and strong functional performance. However, growing concerns regarding their long-term safety and negative consumer perception have prompted the industry to seek naturally derived alternatives. Natural additives sourced from rosemary, stevia, or lactobacillus-fermented products are increasingly gaining acceptance due to their perceived safety and health benefits, although they often pose challenges such as reduced shelf life or formulation instability [69]. In contrast, natural additives from rosemary, stevia, or fermented products are generally well-received, even though they may have shorter shelf life or require more advanced formulation technologies.

Achieving clean-label compliance while maintaining product quality remains one of the most significant challenges for manufacturers. Natural emulsifiers, for instance, typically lack the emulsification strength of synthetic ones, while fermentation-based preservatives can introduce variability in flavor profiles. Similarly, natural sweeteners such as stevia often impart bitter or metallic aftertastes, requiring the use of masking agents to maintain palatability[70].

Despite these challenges, emerging technologies like nanoencapsulation, enzymatic modification, and artificial intelligence (AI)-driven flavor modeling are helping bridge the functional gap between natural and synthetic ingredients. These technological advances are enabling manufacturers to create clean-label products that meet both sensory and performance expectations [71].

Regional variations further illustrate the global evolution of the clean-label movement. In Europe, multinational corporations such as Nestlé and Unilever have taken a leading role by reformulating up to 80% of their food portfolios to meet clean-label standards by 2023 [72, 73]. In contrast, innovation in South Asia has been largely resource-driven, with startups in Bangladesh and India developing cost-effective and biodegradable starch-based emulsifiers derived from jackfruit and tamarind seeds [74]. Meanwhile, in the United States, companies like Chobani and Amy's Kitchen have successfully positioned themselves in the market through strong branding and consumer trust in their additive-free and natural product lines [75, 76]. Collectively, these regional trends demonstrate that clean-label innovation flourishes when it aligns with local resources, consumer awareness, and supportive regulatory environments.

Consumer education and trust remain central to the success of next-generation food additives. Even the most advanced natural additive can face market rejection if consumers do not understand its source or function. For example, terms such as "fermentation-derived nisin" may appear unfamiliar or confusing to the average shopper unless clearly explained. Therefore, transparent labeling, public awareness initiatives, and third-party certifications such as Non-GMO, FOSHU (Foods for Specified Health Uses), and USDA Organic are critical tools for building consumer confidence and credibility in clean-label claims [77, 78].

Finally, environmental sustainability has emerged as an inseparable dimension of additive innovation. The upcycling of food waste – such as using citrus peels as natural antioxidants – not only provides a sustainable raw material source but also helps address global food waste challenges. Similarly, algae and seaweed are being explored as eco-friendly sources of natural salt substitutes, fibers, and stabilizing agents. As sustainability becomes increasingly integrated into food technology, next-generation additives must fulfill dual objectives: promoting human health and ensuring environmental responsibility. This convergence of science, sustainability, and consumer awareness marks a defining era for the future of clean and functional food systems.

9. Conclusion

This chapter explored how modern food additives are changing with the rise of health, transparency, and sustainability demands. Innovations such as plant-based antioxidants, probiotics, nanoencapsulation, and many other advance-based preservation methods are shaping a new generation of additives that are safer, cleaner, and more effective. Although technical and regulatory challenges remain, the shift toward clean-label and health-focused foods continues to grow. Today's consumers want food that not only nourishes but also supports overall well-being and reflects care for the environment. The future of food additives will rely on collaboration among scientists, food technologists, and data experts to develop products that are functional, safe, and sustainable. In the coming years, additives are expected to be more personalized, made from upcycled or eco-friendly sources. Overall, this transformation is not only scientific but also ethical – it reflects a new way of thinking about food and responsibility. By combining innovation with honesty and sustainability, the food industry can create additives that improve health while protecting the planet.

Acknowledgments

The author acknowledges the use of Grammarly for language polishing of the manuscript.

Conflict of Interest

The authors declare no conflict of interest.

Funding

This research did not receive any specific grant from funding agencies in the public, commercial, or not-for-profit sectors.

Author details

Shahnaz Parvin Sweety[1,2*] and Tahsin Ahmed Rupok[1,3]

1 Department of Pharmacy, University of Rajshahi, Rajshahi, Bangladesh

2 Department of Pharmacy, Manarat International University, Dhaka, Bangladesh

3 Department of Pharmacy, World University of Bangladesh, Dhaka, Bangladesh

*Address all correspondence to: shahnazparvin132@gmail.com

References

[1] Aschemann-Witzel J, Varela P, Peschel AO. Consumers' categorization of food ingredients: Do consumers perceive them as 'clean label' producers expect? An exploration with projective mapping. Food Quality and Preference. 2019;**71**:117–128. DOI: 10.1016/j.foodqual.2018.06.003

[2] Chang MY, Chen HS. Understanding Consumers' Intentions to purchase clean label products: Evidence from Taiwan. Nutrients. 2022;**14**(18). DOI: 10.3390/nu14183684

[3] Sarker A. A review on the application of bioactive peptides as preservatives and functional ingredients in food model systems. Journal of Food Processing and Preservation. 2022;**46**(8):e16800. DOI: 10.1111/JFPP.16800;JOURNAL:JOURNAL:17454549;ISSUE:ISSUE:DOI

[4] "Global Clean Label Trends: Consumer insights and preferences" [Online]. Available: https://www.innovamarketinsights.com/trends/global-clean-label-trends-consumer-trends-preferences/ [Accessed: 2025-November-19]

[5] "Mintel announces Global Food and Drink Trends for 2024 | Mintel" [Online]. Available: https://www.mintel.com/press-centre/mintel-announces-global-food-and-drink-trends-for-2024/ [Accessed: 2025-November-19]

[6] Amiri S, Moghanjougi ZM, Bari MR, Khaneghah AM. Natural protective agents and their applications as bio-preservatives in the food industry: An overview of current and future applications. Codon Publications. 2021;**1**. DOI: 10.15586/ijfs.v33iSP1.2045

[7] Teshome E, Forsido SF, Rupasinghe HPV, Keyata EO. Potentials of Natural preservatives to enhance food safety and shelf life: A Review. Hindawi Limited. 2022. DOI: 10.1155/2022/9901018

[8] McClements DJ, Bai L, Chung C. Recent advances in the utilization of natural emulsifiers to form and stabilize emulsions. Annual Reviews Inc. 2017;**28**. DOI: 10.1146/annurev-food-030216-030154

[9] Zhang X, Wang Q, Liu Z, Zhi L, Jiao B, Hu H, Ma X, Agyei D, Shi A. Plant protein-based emulsifiers: Mechanisms, techniques for emulsification enhancement and applications. Food Hydrocolloids. 2023;**144**:109008

[10] Pérez-Díaz IM, Altuntas EG, Juneja VK. Microbial Fermentation in Food Preservation. In: Microbial Control and Food Preservation. New York: Springer; 2017. p. 281–298. DOI: 10.1007/978-1-4939-7556-3_13

[11] Wazir M, Verma H, Singh J, Singh P, Passey S. The Battle of Natural Sweeteners: A comprehensive guide to Monk Fruit and Stevia. Enviro Research Publishers. 2025. DOI: 10.12944/CRNFSJ.13.1.2

[12] Hernandez AV, et al. Effect of low-sodium salt substitutes on blood pressure, detected hypertension, stroke and mortality. Heart. 2019;**105**(12):953–960. DOI: 10.1136/heartjnl-2018-314036

[13] Slavin J. Fiber and prebiotics: Mechanisms and health benefits. MDPI AG. 2013. DOI: 10.3390/nu5041417

[14] Gibson GR, et al. Expert consensus document: The International Scientific Association for Probiotics and Prebiotics (ISAPP) consensus statement on the definition and scope of prebiotics. Nature Publishing Group. 2017. DOI: 10.1038/nrgastro.2017.75

[15] Assadpour E, Jafari SM. A systematic review on nanoencapsulation of food bioactive ingredients and nutraceuticals by various nanocarriers. Taylor and Francis Inc. 2019. DOI: 10.1080/10408398.2018.1484687

[16] Jayaweera Y, Undugoda L, Sandaruwan C, Priyadarshana G. Nanotechnology trends in food processing and food packaging. Nanoscience & Nanotechnology-Asia. 2023;**13**(1). DOI: 10.2174/2210681213666230227111805

[17] Grand View Research. "Clean Label Ingredients Market Size, Share & Trends Analysis Report (Clean Label Ingredients Market Summary)." Available from: https://www.grandviewresearch.com/industry-analysis/clean-label-ingredients-market-report

[18] Asioli D, et al. Making sense of the 'clean label' trends: A review of consumer food choice behavior and discussion of industry implications. Elsevier Ltd. 2017. DOI: 10.1016/j.foodres.2017.07.022

[19] Antich Ferrer E, Fernández-Pastor S, Guerrero A. Impact of food allergies on the Food Safety and Life Quality of Adults in Spain. Foods. 2025;**14**(6):939. DOI: 10.3390/foods14060939

[20] Soutjis B. The new digital face of the consumerist mediator: The case of the 'Yuka' mobile app. Journal of Cultural Economy. 2020;**13**(1):114–131. DOI: 10.1080/17530350.2019.1603116

[21] Nabeshima EH, da Rocha Tavares PE, da Silva Corrêa Lemos AL, de Moura SCSR. Emerging ingredients for clean label products and food safety. Instituto de Tecnologia de Alimentos–ITAL. 2024. DOI: 10.1590/1981-6723.016023

[22] Jaenke R, Barzi F, McMahon E, Webster J, Brimblecombe J. Consumer acceptance of reformulated food products: A systematic review and meta-analysis of salt-reduced foods. Critical Reviews in Food Science and Nutrition. 2017;**57**(16):3357–3372. DOI: 10.1080/10408398.2015.1118009

[23] Novais C, et al. Natural Food Colorants and Preservatives: A review, A demand, and A challenge. American Chemical Society. 2022. DOI: 10.1021/acs.jafc.1c07533

[24] Lelieveld H, Keener L. Global harmonization of food regulations and legislation-the Global Harmonization Initiative. Trends in Food Science & Technology. 2007;**18**(SUPPL 1):S15–S19. DOI: 10.1016/j.tifs.2006.08.005

[25] Meziane H, et al. Rosmarinus officinalis Linn.: Unveiling its multifaceted nature in nutrition, diverse applications, and advanced extraction methods. Springer. 2025. DOI: 10.1007/s43994-024-00144-y

[26] Khan N, Mukhtar H. Tea polyphenols in promotion of human health. MDPI AG. 2019. DOI: 10.3390/nu11010039

[27] Kamah F, et al. Phenolic compounds and biological activities of grape (Vitis vinifera L.) seeds at different ripening stages: Insights from

Algerian varieties. Arquivo Brasileiro de Medicina Veterinária E Zootecnia. 2025;**77**(1). DOI: 10.1590/1678-4162-13361

[28] Aggarwal BB, Harikumar KB. Potential therapeutic effects of curcumin, the anti-inflammatory agent, against neurodegenerative, cardiovascular, pulmonary, metabolic, autoimmune and neoplastic diseases. The International Journal of Biochemistry & Cell Biology. 2009;**41**:40–59. DOI: 10.1016/j.biocel.2008.06.010

[29] Shankar A, Ali A, Abdullah HM, Balaji J, Kaur J, Saeed F, Wasiq M, Imran A, Jibraeel H, Raheem MS, Aslam A. Nutritional composition, phytochemical profile, therapeutic potentials, and food applications of rosemary: A comprehensive review. Journal of Food Composition and Analysis. 2024;**135**:106688

[30] Hegedüs I, et al. The utilization of physiologically active molecular components of Grape Seeds and Grape Marc. MDPI. 2022. DOI: 10.3390/ijms231911165

[31] Aggarwal B, Bharti AC. "Anticancer Potential of Curcumin: Preclinical and Clinical Studies." Available from: https://www.researchgate.net/publication/10818418

[32] Thy LTM, Duy HK, Dat NM. Applications of lecithin in emulsion stabilization and advanced delivery systems in cosmetics: A mini-review. Elsevier B.V. 2025. DOI: 10.1016/j.rsurfi.2025.100543

[33] Brütsch L, Stringer FJ, Kuster S, Windhab EJ, Fischer P. Chia seed mucilage - A vegan thickener: Isolation, tailoring viscoelasticity and rehydration. Food & Function. 2019;**10**(8):4854–4860. DOI: 10.1039/c8fo00173a

[34] Onyango E. "Legume Protein: Properties and Extraction for Food Applications." 2022. Available from: www.intechopen.com

[35] Schreiner TB, Dias MM, Barreiro MF, Pinho SP. Saponins as Natural Emulsifiers for Nanoemulsions. American Chemical Society. 2022. DOI: 10.1021/acs.jafc.1c07893

[36] Leroy F, De Vuyst L. Lactic acid bacteria as functional starter cultures for the food fermentation industry. Trends in Food Science & Technology. 2004;**15**(2):67–78. DOI: 10.1016/j.tifs.2003.09.004

[37] Gharsallaoui A, Oulahal N, Joly C, Degraeve P. Nisin as a Food Preservative: Part 1: Physicochemical properties, antimicrobial activity, and main uses. Critical Reviews in Food Science and Nutrition. 2016;**56**(8):1262–1274. DOI: 10.1080/10408398.2013.763765

[38] Murooka Y. Acetic acid bacteria in production of vinegars and traditional fermented foods. In: Acetic Acid Bacteria: Ecology and Physiology. Japan: Springer; 2016. p. 51–72. DOI: 10.1007/978-4-431-55933-7_2

[39] de Arauz LJ, Jozala AF, Mazzola PG, Penna TCV. Nisin biotechnological production and application: A review. Trends in Food Science & Technology. 2009;**20**:146–154. DOI: 10.1016/j.tifs.2009.01.056

[40] Meena M, Prajapati P, Ravichandran C, Sehrawat R. Natamycin: A natural preservative for food applications—A review.

The Korean Society of Food Science and Technology. 2021. DOI: 10.1007/s10068-021-00981-1

[41] Hill C, et al. Expert consensus document: The international scientific association for probiotics and prebiotics consensus statement on the scope and appropriate use of the term probiotic. Nature Reviews Gastroenterology & Hepatology. 2014;**11**(8):506–514. DOI: 10.1038/nrgastro.2014.66

[42] Ajami M, et al. Effects of stevia on glycemic and lipid profile of type 2 diabetic patients: A randomized controlled trial. Avicenna Journal of Phytomedicine. 2020;**10**(2):118. https://pmc.ncbi.nlm.nih.gov/articles/PMC7103435/

[43] "Monk Fruit: 4 Benefits. Nutrition, and Risks" Available: https://www.health.com/monk-fruit-benefits-7494675 [Accessed: 2025-November-19]

[44] Fda. "The Declaration of Allulose and Calories from Allulose on Nutrition and Supplement Facts Labels: Guidance for Industry", [Online]. Available from: https://www.fda.gov/FoodGuidances [Accessed: 2025-October-22]

[45] He FJ, Tan M, Ma Y, MacGregor GA. Salt Reduction to Prevent Hypertension and Cardiovascular Disease: JACC State-of-the-Art Review. Elsevier USA. 2020. DOI: 10.1016/j.jacc.2019.11.055

[46] Mahadevan K. Seaweeds: A sustainable food source. In: Seaweed Sustainability: Food and Non-Food Applications. Elsevier Inc; 2015. p. 347–364. DOI: 10.1016/B978-0-12-418697-2.00013-1

[47] Şen Yılmaz EB. Utilization of Yeast Extract as a Flavor Enhancer and Masking Agent in Sodium-Reduced Marinated Shrimp. Molecules. 2024;**29**(1). DOI: 10.3390/molecules29010182

[48] Kanekanian A. Encapsulation Technologies and Delivery Systems for Food Ingredients and Nutraceuticals, edited by N. Garti and D.J. McClements, Woodhead Publishing Ltd, Cambridge, UK. International Journal of Dairy Technology. 2013;**66**(3):458–459. DOI: 10.1111/1471-0307.12059

[49] Ramalho MJ, Coelho MAN, Pereira MC. Nanoparticles for delivery of Vitamin D: Challenges and opportunities. In: A Critical Evaluation of Vitamin D-Clinical Overview. InTech; 2017. DOI: 10.5772/64516

[50] Venugopalan VK, et al. Encapsulation and protection of omega-3-rich fish oils using food-grade delivery systems. MDPI AG. 2021. DOI: 10.3390/foods10071566

[51] Chen J, et al. Fabrication and evaluation of curcumin-loaded nanoparticles based on solid lipid as a new type of colloidal drug delivery system. Indian Journal of Pharmaceutical Sciences. 2013;**75**(2):178. DOI: 10.4103/0250-474X.115463

[52] Wang X, et al. Microencapsulating alginate-based polymers for probiotics delivery systems and their application. Pharmaceuticals. 2022;**15**(5):644. DOI: 10.3390/ph15050644

[53] Kumah EA, Fopa RD, Harati S, Boadu P, Zohoori FV, Pak T. Human and environmental impacts of nanoparticles: A scoping review of the current literature. BMC Public Health. 2023;**23**(1). DOI: 10.1186/s12889-023-15958-4

[54] Matsuda R. Biocompatibility and Toxicity Studies of Nanomaterials Understanding Risks and Safety Correspondence to. Journal of Nanomedicine & Nanotechnology. 2024;**15**:753. DOI: 10.35248/2157-7439.24.15.753

[55] More S, et al. Guidance on risk assessment of nanomaterials to be applied in the food and feed chain: Human and animal health. EFSA Journal. 2021;**19**(8). DOI: 10.2903/j.efsa.2021.6768

[56] "Guidance for Industry: Regulatory Framework for Substances Intended for Use in Human Food or Animal Food on the Basis of the Generally Recognized as Safe (GRAS) Provision of the Federal Food, Drug, and Cosmetic Act | FDA." [Online]. Available from: https://www.fda.gov/regulatory-information/search-fda-guidance-documents/guidance-industry-regulatory-framework-substances-intended-use-human-food-or-animal-food-basis [Accessed: 2025-November-20]

[57] Bearth A, Cousin ME, Siegrist M. The consumer's perception of artificial food additives: Influences on acceptance, risk and benefit perceptions. Food Quality and Preference. 2014;**38**:14–23. DOI: 10.1016/j.foodqual.2014.05.008

[58] "Safety evaluation of certain food additives and contaminants: Prepared by the Seventy-Third Meeting of the Joint FAO/WHO Expert Committee on Food Additives (JECFA) [Online]." Available: https://www.who.int/publications/i/item/9789241660648 [Accessed: 2025-November-20]

[59] Güngörmüş C, Kılıç A. The Safety Assessment of Food Additives by Reproductive and Developmental Toxicity Studies. Food Additive. 2012. DOI: 10.5772/30787

[60] FAO/WHO. "Evaluation of certain food additives." World Health Organization (WHO) [Online]. Available from: https://www.who.int/publications/i/item/9789241210294 [Accessed: 2025-September-02]

[61] "Generally Recognized As Safe (GRAS) | FDA" [Online]. Available from: https://www.fda.gov/food/food-ingredients-packaging/generally-recognized-safe-gras [Accessed: 2025-November-20]

[62] Hardy A, et al. Guidance on the risk assessment of substances present in food intended for infants below 16 weeks of age. EFSA Journal. 2017;**15**(5). DOI: 10.2903/j.efsa.2017.4849

[63] FSSAI. "Manual of methods of Analysis of Foods Food Additives Food Safety and Standards Authority of India Ministry of Health and Family Welfare Government of India New Delhi 2016 [Online]." 2016. Available from: https://fssai.gov.in/upload/uploadfiles/files/Manual_Food_Additives_25_05_2016(1).pdf [Accessed: 2025-August-30]

[64] WHO. "Safety evaluation of certain food additives [Online]." 2024. Available from: https://www.who.int/publications/i/item/9789240100978 [Accessed: 2025-August-30]

[65] Kabadi SV, Zang Y, Fisher JW, Smith N, Whiteside C, Aungst J. Food ingredient safety evaluation: Utility and relevance of toxicokinetic methods. Academic Press Inc. 2019. DOI: 10.1016/j.taap.2019.114759

[66] More SJ, et al. Guidance on the use of the benchmark dose approach in risk

assessment. EFSA Journal. 2022;**20**(10). DOI: 10.2903/j.efsa.2022.7584

[67] Brigo A, Müller L. Development of the threshold of toxicological concern concept and its Relationship to Duration of Exposure. Genotoxic Impurities: Strategies for Identification and Control. 2011;27–63. DOI: 10.1002/9780470929377.CH2

[68] "Titanium dioxide E171 no longer considered safe when used as a food additive | EFSA." Available from: https://www.efsa.europa.eu/en/news/titanium-dioxide-e171-no-longer-considered-safe-when-used-food-additive [Accessed: 2025-November-20]

[69] Shahidi F, Ambigaipalan P. Phenolics and polyphenolics in foods, beverages and spices: Antioxidant activity and health effects - A review. Elsevier Ltd. 2015. DOI: 10.1016/j.jff.2015.06.018

[70] Singh K, Tiwari G, And NPS, Upadhyay A. Masking the Bitterness of Stevia rebaudiana Ber. Extract Using Natural Adjuvant and their Effect on Recovery of Steviosides in Final Extract. International Journal of Agriculture Sciences. 2020;**12**(14). https://www.bioinfopublication.org/jouarchive.php?opt=&jouid=BPJ0000217

[71] Chauhan K, Rao A. Clean-label alternatives for food preservation: An emerging trend. Elsevier Ltd. 2024. DOI: 10.1016/j.heliyon.2024.e35815

[72] Global N. "Creating Shared Value and Sustainability Report 2023 Advancing regenerative food systems at scale Contents [Online]." 2023. Available from: www.nestle.com/

[73] Unilever. "A century of Impact, Partnerships and Building a Brighter Future Together." 2023. Available from: https://www.unilever.com/files/92ui5egz/production/a385236bc422513d3c5d65f115177f79728f11a1.pdf [Accessed: 2025-August-07]

[74] Geethalaxmi M, Sunil CK, Venkatachalapathy N. Tamarind seed polysaccharides, proteins, and mucilage: Extraction, modification of properties, and their application in food. Sustainable Food Technology. 2024;**2**(6):1670–1685. DOI: 10.1039/d4fb00224e

[75] Post R. "Learn How Chobani is Leading the Way with Clean Labels," Q1 PRODUCTIONS. [Online]. Available from: https://www.q1productions.com/chobani-leading-the-way-with-clean-labels/ [Accessed: 2025-September-04]

[76] Brown L. "Amy's Kitchen Leads Evolution of Mindful Fast Food." Modern Marketing Partners [Online]. Available from: https://www.modernmarketingpartners.com/2023/12/12/amys-kitchen-leads-evolution-of-mindful-fast-food/ [Accessed: 2025-September-04]

[77] Cao Y, Miao L. Consumer perception of clean food labels. British Food Journal. 2023;**125**(2):433–448. DOI: 10.1108/BFJ-03-2021-0246

[78] Zafar MZ, Shi X, Yang H, Abbas J, Chen J. The impact of interpretive packaged food labels on consumer purchase intention: The comparative analysis of efficacy and inefficiency of food labels. International Journal of Environmental Research and Public Health. 2022;**19**(22):15098. DOI: 10.3390/ijerph192215098

Chapter 2

Natural Sweeteners: Consumer Interest and Safety of Natural Sweeteners

Victoria Mukuni

Abstract

Sweetness is a basic taste that influences food choice and the use of sweeteners by consumers. Modern technology has provided opportunities for the development of new types of sweeteners and the discovery of sugar alternatives. This chapter presents some of the most common natural sweeteners, their current usage around the world, consumer interest in these sweeteners, and their safety. The chapter utilizes existing literature to explore consumer perceptions and interest in natural sweeteners, as well as expert recommendations for their use. Overall, the literature indicates a growing demand for natural sweeteners among consumers with diabetes and other non-communicable diseases, as these individuals seek to regulate their sugar intake. Furthermore, minimal safety risks have been reported regarding the consumption of natural sweeteners. Nevertheless, regulatory bodies provide guidelines on how to properly use these sweeteners by providing acceptable daily intake (ADI) recommendations. Future studies can provide insight into long-term effects and benefits of their use, especially for consumers with sensitivities to nutritive sweeteners and other dietary restrictions.

Keywords: natural sweeteners, safety of natural sweeteners, sugar alternatives, non-caloric sweeteners, consumer interest in natural sweeteners, non-nutritive sweeteners

1. Introduction

Natural sweeteners are compounds that occur naturally and can be used for this purpose. They come in various forms, such as syrups or powders. Sweeteners, in general, have gained popularity due to their low caloric content [1]. The idea of what is considered natural is largely contested and lacks a singular definition in the literature. Furthermore, what is considered natural in one part of the world is not the same in another [2]. Despite the lack of definition, the literature still shows that consumers appreciate products that are labeled natural compared to those that are artificial. Noting what is considered natural provides insight into product formulation and consumer perceptions of such products [3]. Consumers desire products that they believe are minimally processed and healthy [4].

Sweeteners and sugars have multiple functions in food. For example, they act as binding agents, stabilizers, colorants, and more. The most common type of sweetener found in foods is sucrose, a naturally occurring carbohydrate. Sucrose provides energy and sweetness in foods and is therefore considered a nutritive or caloric sweetener due to its chemical structure as a carbohydrate. These sugars can be further categorized as monosaccharides, disaccharides, oligosaccharides, and polysaccharides, which are different types of carbohydrates [5]. Not everything that tastes sweet is classified as a sweetener; it is the chemical structure or composition that determines what is considered a sweetener. Another category of sweeteners includes sugar alcohols or polyols. These are naturally occurring in fruits and vegetables and are typically synthesized through the hydrogenation of carbohydrates. Sugar alcohols tend to be poorly absorbed by the body, even though they provide fewer calories compared to other categories of sweeteners. Consuming these sweeteners in high amounts can have laxative effects [5].

Artificial sweeteners have become increasingly popular among consumers and food industry manufacturers. They can be used to sweeten foods, coat medicines, and serve other purposes. Artificial sweeteners are created through chemical processes. They were first introduced to the United States in the 1950s and have since grown in use and interest among US consumers [6]. As their name suggests, artificial sweeteners are man-made and differ from natural sweeteners in several ways. Artificial sweeteners tend to be much sweeter than caloric and/or nutritive sweeteners, such as sucrose. Due to the role of sweeteners in foods, artificial sweeteners are often used alongside other ingredients, like fillers and anticaking agents. Some studies [6–8] show that fillers, such as maltodextrin, are responsible for intestinal inflammation and changes to the microbiota in the gut. Regulatory agencies, such as the U.S. Food and Drug Administration (FDA), have provided guidelines on the use of artificial sweeteners and have issued a list of approved artificial sweeteners, which include Aspartame, Acesulfame potassium (Ace-K), Sucralose, Neotame, Advantame, and Saccharin [6].

This chapter specifically focuses on consumer interest and the safety of natural sweeteners. It also explores some of the most common types of natural sweeteners, how they differ from other sweeteners, technological challenges, and expert recommendations for their use.

2. Most common natural sweeteners and their sources

Natural sweeteners have gained popularity among US consumers and have been incorporated by the food industry into product formulations. Other countries have also guided the use of these natural sweeteners for industry personnel by providing approval or safety status, such as "Generally Recognized as Safe (GRAS)," "not approved," or "not specified." **Table 1** provides a summary of some of the most common natural sweeteners, along with their common uses, sweetness levels, and their approval or safety status [1, 6].

2.1 Allulose

Allulose (see **Figure 1**), also known as psicose, is produced through the enzymatic conversion of fructose (a type of sugar often derived from fruits) to allulose in the

Name of a natural sweetener	Source	Types of food commonly used in	Level of sweetness in comparison to sucrose	Approval status	References
Allulose	Enzymatic production of fructose	Dried fruits, seasonings, beverages, and baked goods	70%	Approved in China and Japan. GRAS status in the United States	[9, 10]
Monk fruit	*Siraitia grosvenorii*	Matcha powders, beverages	300 times	Approved in the United States, Canada, Japan, and Australia	[15–18]
Thaumatin	Plant protein (Katemfe fruit)	Chewing gum	2,000 to 3,000 times	Approved in the European Union	[15, 21, 22]
Tagatose	Some fruits and dairy products	Ice cream, cereals	90%	GRAS status in the United States	[11, 14]
Stevia glycosides	*Stevia rebaudiana*	Sodas, water, ice cream, baked goods	Depends on the type of glycoside	Approved in the European Union and the United States	[18, 23, 30]
Trehalose glycyrrhizin	Fungi, algae, yeast, and other plants, *Glycyrrhiza glabra*	Baked goods, cereals, pasta, dairy products, seasonings, confectionery, beverages	45%–50%	GRAS status in the United States, approved in the European Union, Australia, Japan, and Brazil; GRAS status in the United States, approved in Japan, the European Union, and China	[25, 31, 32]

Table 1.
Common natural sweeteners summary table.

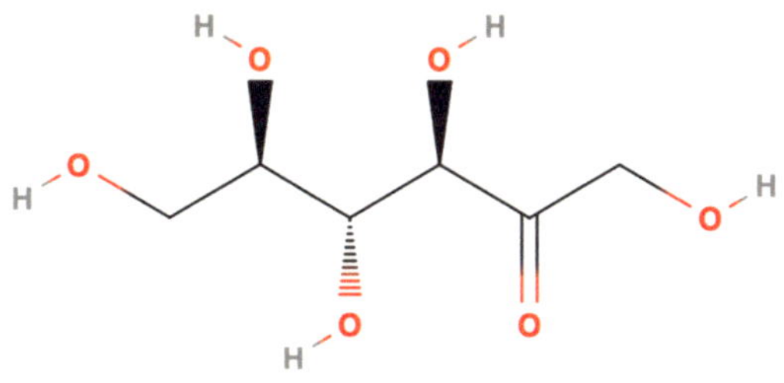

Figure 1.
Chemical structure of allulose.

presence of the enzyme D-Allulose 3-epimerase. In nature, it is found in trace amounts in Itea (a type of flowering shrub), wheat, and some bacteria [9]. Allulose has 70% of the sweetness compared to sucrose (table sugar). Additionally, allulose typically browns faster than table sugar [10].

The low conversion yield of allulose during production negatively impacts the economic and technical yield of the sweetener. During the production of allulose,

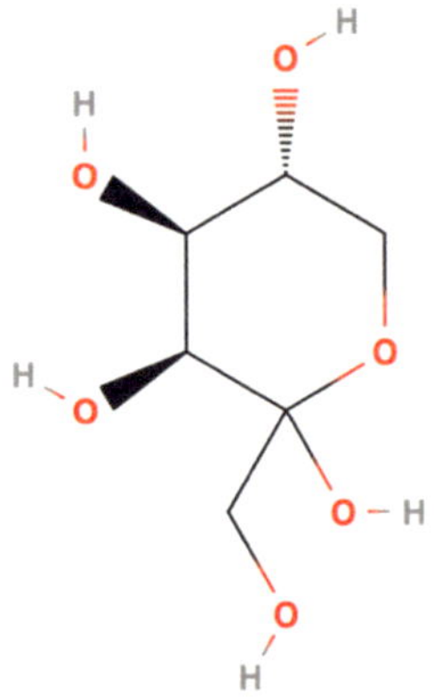

Figure 2.
Chemical structure of D-tagatose.

about 30%–41% yield is attained from the Kiliani-acetonation sequence, a process used to synthesize new sugars [9].

Allulose has several physiological functions, including anti-hyperglycemic effects, anti-hyperlipidemic effects, anti-inflammatory effects, and neuroprotective effects. Some of the health benefits include enhancing insulin resistance, reducing body fat accumulation, and other antidiabetic effects. Allulose promotes gelling in foods and increases flavor while reducing the extent of oxidation during the Maillard browning process [9].

2.2 Tagatose

Tagatose (D-tagatose) is a low-calorie sweetener that is often used as an antidiabetic. Its natural occurrence has been observed in several sources, such as the gum exudates of *Sterculia setigera*, the oligosaccharides of lichen belonging to *Rocella* species, and, finally, in trace amounts in foods such as powdered milk, yogurt, and other dairy products [11]. **Figure 2** shows the chemical structure of D-tagatose.

Tagatose is an excellent Maillard-reactive reducing sugar [12], which is beneficial for manufacturers and consumers preparing browned products, as it adds a rich brown color that is often observed in baked foods such as cookies, seared meats, or toasted bread. Tagatose also enhances flavor by producing flavor notes that are nutty or caramel-like [10]. However, excessive browning of tagatose can produce undesirable flavors and impact shelf-life over time [13]. D-tagatose can be chemically or enzymatically produced and has several uses, including as a humectant, a stabilizer, a texturizer, and a sweetener [12]. Tagatose, as a sweetener, also has probiotic benefits [14]. A literature review on the effects of D-tagatose on blood glucose showed that D-tagatose reduced postprandial blood glucose rise and glycosylated hemoglobin in patients with type 2 diabetes [15].

2.3 Monk fruit

Monk fruit (*Siraitia grosvenorii)* is a perennial herb cultivated in the Guangxi province of China [16, 17]. Monk fruit has a sweetness level of about 300 times that of sucrose. The sweetness of monk fruit comes from mogrosides, which have mogrol backbones and glucose units attached as glycosides [18], as illustrated in **Figure 3**.

Compared to other sweeteners, the availability of monk fruit is not as abundant, which impacts how it is processed for distribution to interested markets. Monk fruit extract has antioxidant and anti-inflammatory properties [17]. There is little evidence that monk fruit affects the gut microbiome. One study found that Mogroside V in monk fruit may have prebiotic potential, promoting the growth of beneficial bacteria in the digestive tract [19].

The stability of monk fruit extract is based on the type of extract and the pH of the storage environment. For example, Monk fruit-25%, Monk fruit-40%, Monk fruit-45%, and Monk fruit-50% were observed to remain stable at 25°C for six months, while Mogroside V in Monk fruit-25% was unstable at pH 1 and at 2°C–8°C [20].

Monk fruit extracts are not recognized by the body as carbohydrates, which means they do not trigger blood sugar or insulin release. A 2025 study found that monk fruit extract can reduce postprandial blood glucose levels by 10% to 18% and insulin response by 12% to 22% in humans [19].

2.4 Thaumatin

Thaumatin is a natural plant protein composed of Thaumatin I and Thaumatin II proteins, along with other plant constituents that often come from acidic, aqueous extractions of the aerial parts of the Katemfe fruit, native to the African plant *Thaumatococcus daniellii* [21]. Thaumatin is 2,000 to 3,000 times sweeter than sucrose [22].

Khayata et al. [15] found that thaumatin, when used for four consecutive weeks at half the average daily intake of sugar, did not cause significant effects on the short- or long-term glycemic response in individuals without diabetes. However, in individuals with type 2 diabetes, thaumatin indirectly decreased blood glucose significantly. The researchers concluded that thaumatin is a safe, natural sweetener that can benefit patients with type 2 diabetes. **Figure 4** shows the chemical structure of thaumatin.

2.5 Stevia glycosides

Stevia glycosides are a group of plant-derived compounds that are often used in foods and beverages as sweeteners. *Stevia Rebaudiana* is a perennial shrub that is indigenous to South America. Indigenous peoples have long used the leaves of the Stevia plant to sweeten their foods. Stevia glycosides have been found to exhibit very potent antidiabetic activity [1, 23, 24]. **Figure 5** illustrates the chemical structure of stevioside, a major Stevia glycoside.

2.5.1 Trehalose

Trehalose, a naturally occurring disaccharide, is composed of two glucose molecules, as shown in **Figure 6**. It is commonly found in plants, animals, insects, fungi, and bacteria. Due to the chemical structure of trehalose, the sugar does not brown in Maillard reactions because of the absence of free aldehyde groups that are often present in reduction reactions [25]. Trehalose shows limited impact on glycemic indices [26].

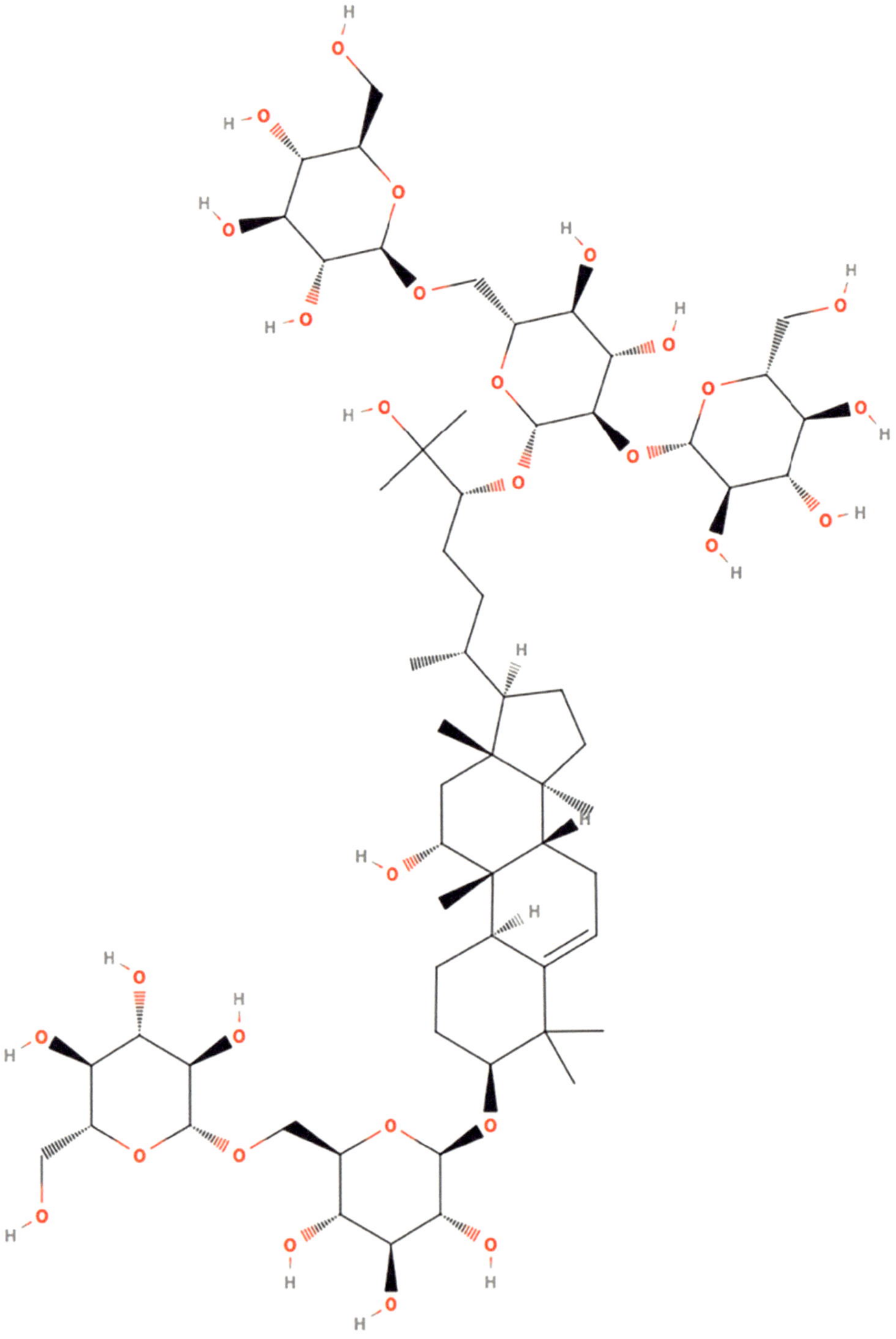

Figure 3.
Chemical structure of mogroside V.

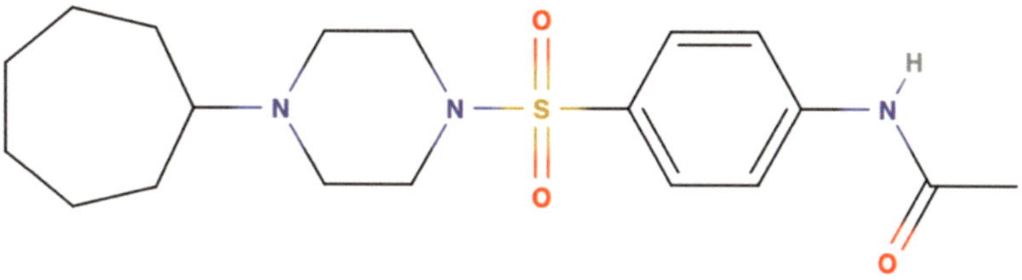

Figure 4.
Chemical structure of thaumatin.

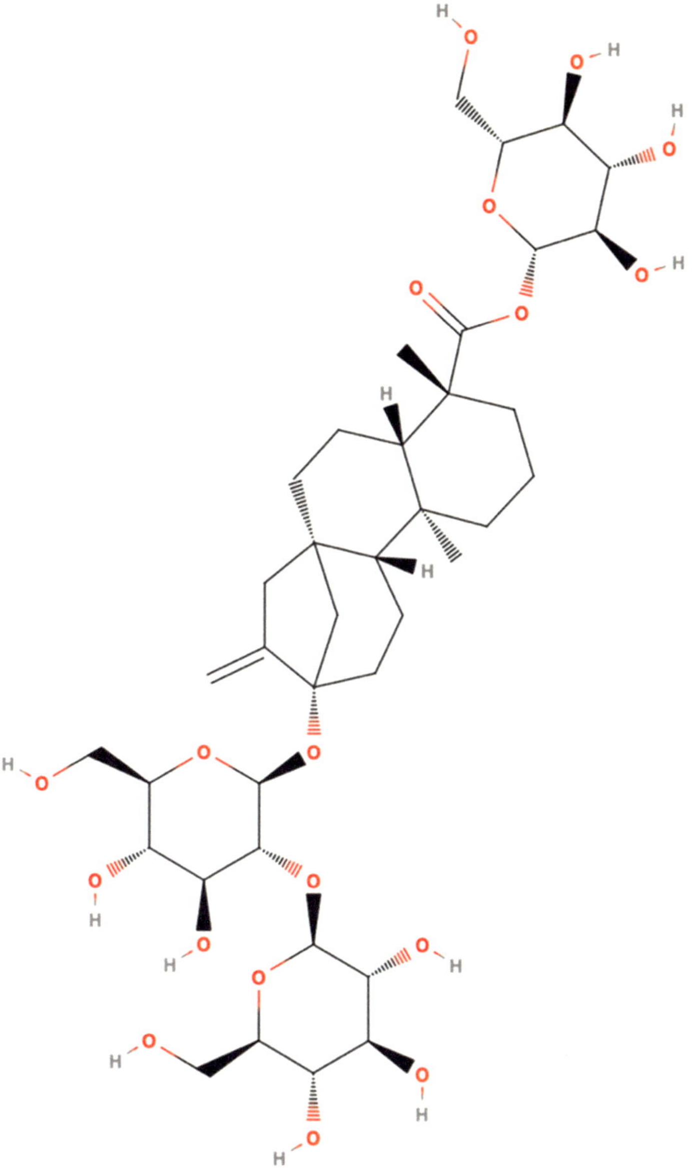

Figure 5.
Chemical structure of stevioside, a major stevia glycoside.

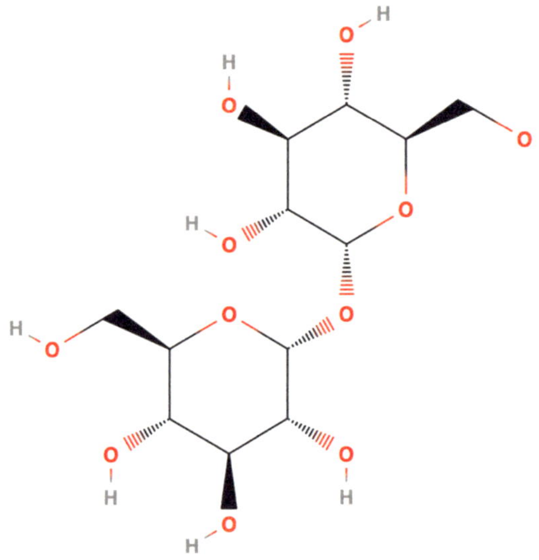

Figure 6.
Chemical structure of trehalose.

2.5.2 Glycyrrhizin

Glycyrrhizin, also known as licorice, is derived from the licorice plant, *Glycyrrhiza glabra* L. As shown in **Figure** 7, glycyrrhizin is a triterpene glycoside, joined to a disaccharide of glycyrrhizic acid [27].

Glycyrrhizin is said to have antioxidant, anti-inflammatory, and immunosuppressive properties [28] and is also used for other purposes due to its medicinal properties, such as in cough syrup and in the treatment of stomach ulcers and constipation. With a sweetness level of 30–50 times that of sucrose, glycyrrhizin has no impact on the glycemic index. No acceptable daily intake (ADI) is currently established for it; however, the European Union has suggested a limit of 100 mg per day [29].

Table 1 provides a list of seven common natural sweeteners, their sources, the most common types of foods they are found in, their level of sweetness in comparison to sucrose, and their approval or safety designation in several countries and regions.

2.6 Classification of natural sweeteners as additives

Natural sweeteners are termed as such due to their source [33]. They can be further classified into other categories, such as "nutritive, including polyols, rare sugars, honey, maple syrup, agave, and nonnutritive, which include steviol glycosides, rebaudiosides, Luo Han Guo (Monk fruit), and thaumatin" [33]. Non-nutritive sweeteners (NNS) are usually used in place of refined sugar and can be sourced either naturally or artificially. Natural sweeteners tend to exhibit diverse physicochemical properties, which determine how they can be used. These properties can range from solubility and color to pH [34].

As food additives, natural sweeteners must be approved by authoritative organizations, such as the Food and Agriculture Organization (FAO), the Joint Expert Committee on Food Additives (JECFA), the European Food Safety Authority,

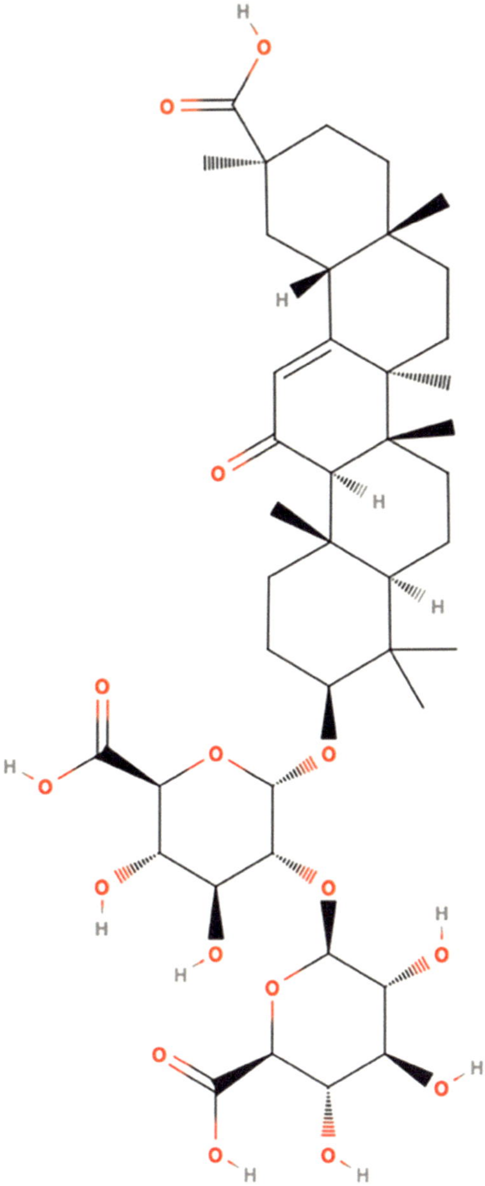

Figure 7.
Chemical structure of glycyrrhizin.

the Food and Drug Administration (FDA), and other region or country-specific organizations. These organizations prescribe ADIs for each type of natural sweetener, or the amount that is acceptable with minimal risk [35].

2.7 Vegan or plant-derived sweeteners

Another category of sweeteners is vegan sweeteners, which are derived from plant sources rather than being synthesized by bacteria. Many natural sweeteners fall under the vegan category and can help reduce blood sugar spikes due to their low glycemic index. Additionally, they offer other health benefits, such as providing vitamins and minerals. These sweeteners are particularly beneficial for consumers who prefer vegan options. Some commercially available vegan sweeteners in syrup form include Manuka honey, yacon root syrup, maple syrup, brown rice syrup, date syrup, agave nectar, corn syrup, and palm syrup [36].

3. Food safety considerations of natural sweeteners

Food safety, like any edible product, is affected by factors such as storage conditions and methods of consumption. Food safety describes the scientific and methodological approaches that are used to ensure the safety of food and to prevent foodborne illnesses or any other adverse impacts on health [37]. Most regulatory authorities have established that there are minimal to no dangers associated with the consumption of natural sweeteners when they are consumed within the specified limits, that is, their ADIs [see 20, 22, 38].

While some sweeteners are approved for use in certain countries, they remain unapproved in others and are still under evaluation for consumer safety. In the United States, for example, several natural sweeteners have received the categorical label of "Generally Recognized as Safe" or GRAS to signal to food manufacturers (and consumers) that the sweetener can be used in product formulations pending additional research [39].

Consumers tend to have a positive outlook on natural sweeteners in foods. The less processed the food is, the better it is for a growing number of consumers around the world [40]. Other aspects are coming into play when it comes to foodstuffs, including the price and taste of a food. Safety concerns in the production or processing of natural sweeteners have been the number one hindrance in the move toward natural sweeteners in the food industry [40].

Many artificial sweeteners have been linked to causing adverse impacts on health, such as disruptions to the gut microbiome, laxative effects, and, more seriously, increasing the risk of carcinogenic effects [41, 42]. For consumers, these adverse effects from the consumption of artificial sweeteners may be translated to natural sweeteners and could affect market saturation due to negative perceptions of sugar alternatives. It is, therefore, significant for industry professionals to advocate for natural sweeteners, especially relying on the fact that these sweeteners are produced naturally and not artificially. Studies conducted to understand consumer perception and preference for natural versus artificial sweeteners show that "consumers have received natural sweeteners positively, especially consumers who use these sweeteners for health purposes. Consumers are eager to purchase products with natural ingredients and clean labels, preferably with further functional properties, but which do not compromise taste. To achieve this trend, food industries are now willing to reformulate their food products to include alternative natural sweeteners to sugar" [40].

From a safety perspective, several factors should be considered in understanding how consumers use sweeteners. Considerations such as how the sweeteners are stored, their shelf life or stability in different environments, consumption amounts, the individual consumer's tolerance to sweeteners, the water activity [50], and other factors play an important role in determining how

Natural sweetener	Amount recognized as safe	References
Allulose	Not specified	
Monk fruit	Not specified	
Thaumatin	Not specified	
Tagatose	45 g/day or 0.75 g/kg body weight/day	[11]
Stevia glycosides	4 mg/kg body weight	[38]
Trehalose	Not specified	
Glycyrrhizin	Not specified	

Table 2.
Recommended intake of some natural sweeteners. Many natural sweeteners have been recognized as safe due to the body's ability to absorb rather than digest these natural sweeteners [20].

safe they are. It is not merely the product formulation but also how the consumer uses the sweetener that determines the level of safety. However, the burden of safety is also on the manufacturers to utilize clear labeling on how to properly store these sweeteners and how to use them to prevent adverse impacts on health. Some consumers are sensitive or intolerant to certain types of sugars. This is due to the body's inability to break down certain types of sugars or the maldigestion of carbohydrates [43]. Intolerances to table sugar, for example, can cause symptoms such as nausea, vomiting, bloating, and stomach pain [44]. Food manufacturers must be transparent about their formulations by listing the type of sweetener or sugar that is used in a product. It is therefore a safety concern for consumers and an important factor for manufacturers to list ingredients, along with other products that have been processed in the same facility as these sweeteners. **Table 2** provides a list of common natural sweeteners and the recommended daily intake and/or the amount recognized as safe.

Generally, there are minimal adverse impacts on health from natural sweeteners. However, even specified amounts can cause minimal health issues. Tagatose, for example, is a malabsorbed sugar in the small intestine due to its structure [11]. This presents a considerable food safety risk for consumers, although it does not make the sugar dangerous for consumption. It also means that greater regulation is warranted on how tagatose is used in foods to protect consumers from the undue impacts of the sweetener on their small intestine and gut health in general. **Table 2** shows information that should be readily available to consumers to mitigate the risk of gut inflammation due to natural sweeteners like tagatose, as growing interest, demand, and use of natural sweeteners in foods continue to increase [40].

4. Expert recommendations on the use of natural sweeteners

Organizations such as the World Health Organization (WHO) and country-specific organizations advise the use of natural sweeteners up to a particular level. The WHO recognizes the need to reduce obesity and body mass index, which have increased due to the growing number of people who are overweight around the world. Furthermore, the WHO states that noncommunicable diseases, such as cardiovascular diseases and type 2 diabetes, are some of the risk factors that are heightened due to obesity. High levels of free sugar intake are often associated with poor dietary quality. This has emphasized the importance of consuming sugar-free foods and non-sugar sweeteners to mitigate these risk factors [38].

The WHO published conditional recommendations for using non-sugar sweeteners as a means of achieving weight control or reducing noncommunicable diseases. The WHO includes several artificial sweeteners and natural sweeteners such as Stevia and Stevia extract. These recommendations were based on systematic reviews conducted to assess the overall health effects of consuming non-sugar sweeteners in individuals, including pregnant women [38].

Overall, the literature shows more statements by authoritative and scientific advisory boards on nonnutritive sweeteners of artificial origin than on natural sweeteners. The American Heart Association, for example, in 2012 made statements advising consumers on NNSs from both natural and artificial sources. Many times, in the literature, these NNSs are permissible, albeit through serving suggestions [42].

The International Sweeteners Association (ISA) states that sweeteners are some of the most intensively researched ingredients, and their safety has been confirmed by global food safety organizations, including the European Food Safety Authority, the U.S. FDA, and the JECFA of the FAO/WHO. The association asserts that the consumption of approved sweeteners is safe for all population groups based on extensive research [30, 45].

Regarding low- or no-calorie sweeteners, the association provides guidance for consumers in European countries on how to identify these sweeteners in food products. Food labels typically include the term "sweetener" along with an E-number, which indicates that the ingredient or additive is considered safe within Europe. To reassure consumers about the safety of no-calorie sweeteners, the association notes that certain NNSs, like aspartame, undergo reevaluation as part of a scientific risk assessment process to reconfirm their safety [30, 45].

5. Technological challenges and benefits of using natural sweeteners

Food industry professionals consider the following properties and others when it comes to using natural sweeteners in foods, including the sweetener's ability to withstand diverse gradients and formulations, stability at different temperatures, lack of carcinogenic effects, high yield in the production process, and affordability, along with other properties. This is because replacing sucrose with a natural sweetener contributes to more than just sweetness in a product. Sucrose provides mouthfeel, bulking, freezing point depression, browning, antioxidation (in meats), curdling of milk, aroma, texture, color, and viscosity [45, 46]. Replicating the multifunctionality of sucrose presents several technological challenges in food formulations that use natural sweeteners. For example, while stevia glycosides, monk fruit extracts, and other sugar alcohols have been approved by regulatory agencies, there are still technological challenges associated with their use in food formulations. These sweeteners often cannot be used alone in food products and typically require a combination with other ingredients. For instance, stevia glycosides are known to have a bitter, licorice-like taste and a metallic aftertaste, particularly at high concentrations and in acidic combinations. Similarly, monk fruit extract may leave a persistent sweet aftertaste with a mild herbal undertone. Thaumatin, on the contrary, has a delayed onset, resulting in a prolonged sweetness that disrupts the overall flavor profile [47].

Natural sweeteners are used not only in the food industry but also in the pharmaceutical industry to coat medications and pills. In the food industry, natural sweeteners are used in the preparation of desserts, candies, juices, chewing gum, and other food items [48].

With growing interest in sugar-free products, consumer interest in sugar substitutes is increasing [40]. Furthermore, the health benefits of sugar-free products have provided incentives for consumers to desire sugar-free products, especially consumers with diabetes and other noncommunicable diseases. However, consumers have expressed concern following previous links between sugar substitutes and cancer. Scientific evidence has dispelled some of these misconceptions through organizations such as the WHO and the U.S. FDA [38].

The Asian market is especially noteworthy, as many natural sweeteners are currently being used and/or have been approved for use by regulators [25, 49]. A major hindrance to the use of natural sweeteners in other parts of the world is the production expenditure of certain sugar substitutes, which have overall lower output during production [9]. The United States, South Korea, and Japan, for example, have seen a viable market for Allulose through its adoption in foods and beverages. China, however, has begun using Allulose as a healthier alternative sweetener in foods, making the market for Allulose greatest in China [49].

Another issue of concern for the market retainability of sugar substitutes is the functionality of natural sweeteners. Sugar, especially sucrose (table sugar), functions as a flavor enhancer, a texture preserver, and provides structural integrity to a food product. Replacing sugar does impact these components of a food, therefore requiring its use alongside other compounds, which can be expensive for food manufacturers [9]. Replacing sucrose in product formulations has technological implications related to functional properties such as bulking, texture, moisture retention, and browning. Not all natural sweeteners are able to replace these properties in foods. For food scientists, achieving desirable structural and physicochemical properties is a challenge, which in turn limits how consumers and manufacturers adopt alternative sweeteners [50].

Moreover, natural sweeteners and other sucrose replacements also affect various technological processes and sensory attributes in baked products. These include starch gelatinization, pasting behavior, and swelling properties. When these attributes are affected, they can impact the overall appearance and acceptability of the final product for consumers and manufacturers [51].

When it comes to sugar alcohols as types of natural sweeteners, they have notable physicochemical effects and cause changes in sensory attributes, especially during storage. One study on how the compositional properties of foods are affected by storage time showed that sensory attributes changed over time, including variations in color, slow degradation rates, and faster decomposition depending on the type of natural sweetener [52]. These physicochemical properties make it difficult for food manufacturers to develop product formulations that extend shelf life while minimizing sensory impacts on the products.

6. Conclusion(s)

Sweetness is one of the most basic flavors in the human diet. With the growing interest in consumer safety, both manufacturers and consumers have sought alternatives to sugary foods, using artificial and natural sweeteners. However, researchers have found that artificial sweeteners may negatively impact consumers' health by increasing the risk of cardiovascular diseases. In contrast, many natural sweeteners are considered to have minimal adverse impacts on health.

This chapter provides an overview of safety concerns regarding natural sweeteners, emphasizing ongoing research to evaluate their potential health impacts, while also discussing some of the most common natural sweeteners and their current usage by consumers and the food industry. Governing bodies around the world have offered guidance for their use by categorizing them as either approved or GRAS, which indicates their safety levels.

Consumers who are sensitive to specific sweeteners may want to know how these sweeteners are labeled, including their ADI values and any other ingredients used to stabilize or alter the texture of natural sweeteners. Therefore, it is important for labels to be accurate and clear to avoid putting consumers with sensitivities at risk. Further studies can explore the long-term effects of natural sweeteners on the health of consumers.

Conflict of Interest

The author(s) declare no conflicts of interest

Author details

Victoria Mukuni
Virginia Tech, Blacksburg, USA

*Address all correspondence to: vicky95@vt.edu

References

[1] Grembecka M. Natural sweeteners in a human diet. Roczniki Panstwowego Zakladu Higieny. 2015;**66**(3):195–202

[2] Baines D. Defining the term 'natural' in the context of food products. In: Baines D, Seal R, editors. Natural Food Additives, Ingredients, and Flavourings. Cambridge: Woodhead Publishing; 2012. p. 1–22. DOI: 10.1533/9780857095725.1

[3] Chambers E, Tran T, Chambers E. Natural: A $75 billion word with no definition—why not? Journal of Sensory Studies. 2019;**34**(4):e12501. DOI: 10.1111/joss.12501

[4] Koppes S. Most Consumers see processed foods as potentially unhealthy but buy them anyway [Internet]. 2025 March 17. Available from: https://ag.purdue.edu/news/2025/03/most-consumers-see-processed-foods-as-potentially-unhealthy-but-buy-them-anyway.html [Accessed: 2025-November]

[5] Zaitoun M, Ghanem M, Harphoush S. Sugars: Types and their functional properties in food and human health. International Journal of Public Health Research. 2018;**6**(4):93–99

[6] Basson AR, Rodriguez-Palacios A, Cominelli F. Artificial sweeteners: History and new concepts on inflammation. Frontiers in Nutrition. 2021;**8**:746247. DOI: 10.3389/fnut.2021.746247

[7] Arnold AR, Chassaing B. Maltodextrin, modern stressor of the intestinal environment. Cellular and Molecular Gastroenterology and Hepatology. 2019;**7**(2):475–476. DOI: 10.1016/j.jcmgh.2018.09.014

[8] Nickerson KP, McDonald C. Crohn's disease–associated adherent-invasive Escherichia coli adhesion is enhanced by exposure to the ubiquitous dietary polysaccharide maltodextrin. PLoS One. 2012;**7**(12):e52132. DOI: 10.1371/journal.pone.0052132

[9] Zhang W, Yu S, Zhang T, Jiang B, Mu W. Recent advances in D-allulose: Physiological functionalities, applications, and biological production. Trends in Food Science & Technology. 2016;**54**:127–137. DOI: 10.1016/j.tifs.2016.06.004

[10] Wang L, Cui Y, Lu Y, Zhao Z. Comprehensive analysis of allulose production: A review and update. Foods. 2024;**13**(16):2572. DOI: 10.3390/foods13162572

[11] Roy S, Chikkerur J, Roy SC, Dhali A, Kolte AP, Sridhar M, et al. Tagatose as a potential nutraceutical: Production, properties, biological roles, and applications. Journal of Food Science. 2018;**83**(11):2699–2709. DOI: 10.1111/1750-3841.14358

[12] Vera C, Guerrero C, Illanes A. Trends in lactose-derived bioactives: Synthesis and purification. Systems Microbiology and Biomanufacturing. 2022;**2**(3):393–412

[13] Grant LD, Bell LN. Physical and chemical stability of tagatose powder. Journal of Food Science. 2012;**77**(3):C308–13. DOI: 10.1111/j.1750-3841.2011.02591.x

[14] Bell LN. Tagatose stability in beverages as impacted by composition and thermal processing. In: Victor R. Preedy, editors. Processing and Impact

on Active Components in Food. London: Academic Press; 2015. p. 613–618. DOI: 10.1016/B978-0-12-404699-3.00074-3

[15] Khayata W, Kamri A, Alsaleh R. The glycemic effect of thaumatin and its mixture with sucrose in type 2 diabetes patients. International Journal of Academic Scientific Research. 2016;**4** (4):38–46

[16] Pandey AK, Chauhan OP. Monk fruit (Siraitia grosvenorii) - health aspects and food applications. Pantnagar Journal of Research. 2019;**17** (3):191–198

[17] Yeung AWK. Bibliometric analysis on the literature of monk fruit extract and mogrosides as sweeteners. Frontiers in Nutrition. 2023;**10**:1253255. DOI: 10.3389/fnut.2023.1253255

[18] Wazir M, Verma H, Singh J, Singh P, Passey S. The battle of natural sweeteners: A comprehensive guide to monk fruit and stevia. Current Research in Nutrition and Food Science Journal. 2025;**13**(1):24–45. DOI: 10.12944/CRNFSJ.13.1.2

[19] Kaim U, Labus K. Monk fruit extract and sustainable health: A PRISMA-guided systematic review of randomized controlled trials. Nutrients. 2025;**17**(9):1433. DOI: 10.3390/nu17091433

[20] EFSA Panel on Food Additives and Flavourings (FAF). Safety of use of monk fruit extract as a food additive in different food categories. EFSA Journal. 2019;**17**(12):e05921. DOI: 10.2903/j.efsa.2019.5921

[21] United States Food and Drug Administration. Thaumatin sweetener and food flavor modifier [Internet]. 2017. Available from: https://www.fda.gov/media/124271/download [Accessed: 2025-November]

[22] EFSA Panel on Food Additives and Flavourings (FAF). Re-evaluation of thaumatin (E 957) as a food additive. EFSA Journal. European Food Safety Authority. 2021;**19**(11):e06884. DOI: 10.2903/j.efsa.2021.6884

[23] Wang J, Zhao H, Wang Y, Lau H, Zhou W, Chen C, Tan S. A review of stevia as a potential healthcare product. Trends in Food Science & Technology. 2020;**103**:264–281. DOI: 10.1016/j.tifs.2020.07.023

[24] Ceunen S, Geuns Jan MC. Steviol glycosides: Chemical diversity, metabolism, and function. Journal of Natural Products. 2013;**76**(6):1201–1228. DOI: 10.1021/np400203b

[25] Shen A, Tapia H, Goddard JM, Gibney PA. Trehalose and its applications in the food industry. Comprehensive Reviews in Food Science and Food Safety. 2022;**21**(6):5004–5037. DOI: 10.1111/1541-4337.13048

[26] Hashemian S, Shojaei M, Radbakhsh S, et al. Effects of oral trehalose on glycaemia, inflammation, and quality of life in patients with type 2 diabetes. Archives of Medical Science: AMS. 2023;**19**(6):1693–1701. DOI: 10.5114/aoms/159048

[27] Fry JC. Natural low-calorie sweeteners. In: Natural Food Additives, Ingredients and Flavourings. Cambridge: Woodhead Publishing; 2012. p. 41–75 DOI: 10.1533/9780857095725.1.41

[28] Seeff L, Stickel F, Navarro VJ. Hepatotoxicity of herbals and dietary supplements. In: Kaplowitz, Neil, DeLeve, Laurie D, editors. Drug-

induced Liver Disease. London: Academic Press; 2013. p. 631–657

[29] Michael Zeece. Chapter Seven - Food additives, Introduction to the Chemistry of Food. Academic Press; 2020. p. 251–311. DOI: 10.1016/B978-0-12-809434-1.00007-4

[30] Tey SL, Salleh NB, Henry J, Forde CG. Effects of aspartame-, monk fruit-, stevia- and sucrose-sweetened beverages on postprandial glucose, insulin and energy intake. International Journal of Obesity. 2017;**41**:450–457. DOI: 10.1038/ijo.2016.225

[31] Agarwal N, Singh SP. A novel trehalose synthase. Microbiology Spectrum. 2021;**9**(3):e01333–21. DOI: 10.1128/Spectrum.01333-21

[32] MS AE-L, El-Razek A, Massoud MI, et al. Utilization of glycyrrhizin and licorice extract. Journal of Food and Dairy Sciences. 2017;**8**(3):127–136. DOI: 10.21608/jfds.2017.37143

[33] Sun Y, Liang J, Zhang Z, et al. Extraction, physicochemical properties, bioactivities and application of natural sweeteners. Food Chemistry. 2024;**457**:140103

[34] Arshad S, Rehman T, Saif S, et al. Replacement of refined sugar by natural sweeteners: Focus on potential health benefits. Heliyon. 2022;**8**(9):e10711. DOI: 10.1016/j.heliyon.2022.e10711

[35] Silva MM, Reboredo FH, Lidon FC. Sweetener food additives: A synoptical overview. Emirates Journal of Food and Agriculture. 2023;**35**:1–6. DOI: 10.9755/ejfa.2023.3202

[36] Anwar S, Syed QA, Munawar F, et al. Inclusive overview of sweeteners trends. ACS Food Science & Technology. 2023;**3**(2):245–255. DOI: 10.1021/acsfoodscitech.2c00325

[37] Food and Agriculture Organization of the United Nations. Food Safety and Quality [internet]. Rome FAO. Available from https://www.fao.org/food-safety/about/q-a-on-food-safety/en [Accessed: 2025-July-11]

[38] World Health Organization. Use of non-sugar sweeteners: WHO guideline summary [Internet]. Institute of Repository for Information; 2023. Available from: https://iris.who.int/server/api/core/bitstreams/da17bb58-541e-4fd8-959c-512c85cf9f5a/content [Accessed: 2025-November]

[39] Świąder K, Wegner K, Piotrowska A, et al. Plants as a source of natural high-intensity sweeteners. Journal of Applied Botany and Food Quality. 2019;**92**:160–168

[40] Saraiva A, Carrascosa C, Raheem D, et al. Natural sweeteners: Food naturalness, sustainability and health impacts. International Journal of Environmental Research and Public Health. 2020;**17**(17):6285. DOI: 10.3390/ijerph17176285

[41] Yan S, Yan F, Liu L, et al. Can artificial sweeteners increase cancer risk? Nutrients. 2022;**14**(18):3742. DOI: 10.3390/nu14183742

[42] Ruiz-Ojeda FJ, Plaza-Díaz J, Sáez-Lara MJ, Gil A. Effects of sweeteners on the gut microbiota. Advances in Nutrition. 2019;**10**(Suppl 1):S31–S48. DOI: 10.1093/advances/nmy037

[43] Fernández-Bañares F. Carbohydrate maldigestion and intolerance. Nutrients. 2022;**14**(9):1923. DOI: 10.3390/nu14091923

[44] Tuck CJ, Biesiekierski JR, Schmid-Grendelmeier P, Pohl D. Food intolerances. Nutrients. 2019;**11**(7):1684. DOI: 10.3390/nu11071684

[45] International Sweeteners Association. What are low/no-calorie sweeteners: [internet]. 2025. Available from: https://www.sweeteners.org/what-are-low-no-calorie-sweeteners/ [Accessed: 2025-July-26]

[46] Cotton RH, Rebers PA, Maudru JE, Rorabaugh GU. The role of sugar in the food industry. Food Technology. Year not available. **12**;3–20

[47] Tahmouzi S, Marand SA, Tarahi M, et al. Application of FDA-approved artificial sweeteners. Comprehensive Reviews in Food Science and Food Safety. 2025;**24**(4):e70208. DOI: 10.1111/1541-4337.70208

[48] Priya K, Gupta VRM, Srikanth K. Natural sweeteners: A complete review. Journal of Pharmaceutical Research. 2011;**4**(7):2034–2039

[49] Liu Z, Li S, Peng J. Exploring the relationship between sugar and sugar substitutes- Analysis of income level and beverage consumption market pattern based on the perspective of healthy China. Nutrients. 2022;**14**(21):4474. DOI: 10.3390/nu14214474

[50] Struck S, Jaros D, Brennan CS, Rohm H. Sugar replacement in sweetened bakery goods. International Journal of Food Science & Technology. 2014;**49**(9):1963–1976. DOI: 10.1111/ijfs.12617

[51] Woodbury TJ, Lust AL, Mauer LJ. The effects of commercially available sweeteners (sucrose and Sucrose replacers) on wheat starch gelatinization and pasting, and cookie baking. Journal of Food Science. 2021;**86**(3):687–698. DOI: 10.1111/1750-3841.15572

[52] Benedek C, Bodor Z, Merrill VT, et al. Effect of sweeteners and storage on blackberry jams. European Food Research and Technology 2020;**246**:2187–2204

Chapter 3

Plant-Derived Antioxidants: Natural Preservation and Health Benefits

Nawab Nasir, Din Mohammad Mushfiq and Sevinç Ateş

Abstract

Plant-derived antioxidants have gained significant attention in recent years due to their dual role in promoting human health and enhancing food preservation. This chapter explores the key natural sources of antioxidants, their mechanisms of action, and their effectiveness in extending shelf life and preventing spoilage in food systems. Emphasis is placed on the health-promoting properties of these compounds, including their role in mitigating oxidative stress, preventing chronic diseases, and supporting immune and cognitive functions. The chapter also examines modern extraction methods – such as ultrasound-assisted and supercritical fluid extraction – that influence antioxidant yield and stability. Regulatory considerations in the USA and EU are discussed, highlighting the importance of truthful labeling and safety standards. Finally, the chapter addresses challenges such as stability, standardization, and large-scale production, while outlining future trends involving biotechnology and nanotechnology. Overall, plant-derived antioxidants represent a sustainable and multifunctional approach to improving food safety and human health.

Keywords: functional foods, natural preservatives, oxidative stress, plant-derived antioxidants, polyphenols

1. Introduction

Plant-derived antioxidants have gained increasing attention for their ability to prevent oxidative damage in both food products and biological systems. Oxygen, though essential for metabolism, can lead to the formation of reactive oxygen species (ROS), reactive nitrogen species (RNS), and various free radicals, which trigger lipid oxidation in foods and oxidative stress in living cells [1–3]. These oxidative processes reduce food quality, promote rancidity, and contribute to the development and progression of chronic diseases such as diabetes, cardiovascular diseases, neurodegenerative disorders, and cancer [1–12]. Plant-derived antioxidants, naturally occurring in fruits and vegetables, neutralize free radicals and interrupt chain reactions involved in oxidative degradation [13]. While synthetic antioxidants, such as butylated hydroxyanisole (BHA) and butylated hydroxytoluene (BHT), have long been used in the food industry, growing safety concerns have led to a shift toward natural alternatives, including vitamin C, vitamin E, lycopene, and various polyphenols [1, 13]. These compounds not only enhance food shelf life but also contribute to disease prevention, highlighting their value in functional foods.

Reactive species arise both exogenously, from environmental factors such as pollution, radiation, and diet, and endogenously, through regular cellular metabolism [2]. When their production exceeds the body's antioxidant defenses, oxidative stress occurs, disrupting cellular signaling and function [2, 4, 5]. To counteract this, organisms have developed a three-line antioxidant defense system: first-line endogenous enzymes such as glutathione peroxidase (GPx), catalase (CAT), and superoxide dismutase (SOD) neutralize radicals directly; second-line low-molecular-weight dietary antioxidants such as glutathione, carotenoids, flavonoids, vitamin C, and vitamin E scavenge radicals and maintain redox homeostasis; and third-line repair/removal systems restore cellular function by eliminating or replacing oxidized biomolecules [2].

In addition to their biological roles, antioxidants are widely employed in the food, nutraceutical, and pharmaceutical sectors to prevent oxidative spoilage, preserve quality, and prolong shelf life [14, 15]. In vitro assays such as 2,2-diphenyl-2-picrylhydrazyl (DPPH), 2,2′-azinobis (3-ethylbenzothiazoline-6-sulfonic acid) (ABTS), ferric reducing antioxidant power (FRAP), cupric ion reducing antioxidant capacity (CUPRAC), and superoxide, hydroxyl, and nitric oxide radical scavenging assays are frequently used to assess antioxidant effectiveness [14, 15]. However, differences in sensitivity and specificity among these methods complicate comparisons across studies.

Despite advances, standardized and universally recognized methods for measuring antioxidant capacity remain lacking, making it challenging to assess effectiveness in complex matrices [14, 15]. Furthermore, while plant-derived antioxidants show strong radical scavenging in vitro, their clinical efficacy in humans is often limited by bioavailability, metabolic transformations, and interactions with other dietary components [2, 3]. More research is also needed to evaluate the safety, delivery methods, and long-term therapeutic potential of enzyme mimics and nanozymes, which remain underutilized [2]. A conceptual gap exists, as many reviews focus narrowly on antioxidant chemistry, biological functions, or measurement techniques without integrating information on natural sources, extraction processes, and quantification instruments, complicating the development of a comprehensive framework for antioxidant science [16].

In food systems, oxidative stress damages lipids, proteins, and other macromolecules, causing rancidity, off-flavors, discoloration, and reduced nutritional value [7]. This accelerates spoilage, particularly in animal-derived foods such as meat and milk. Diets rich in plant-derived antioxidants from fruits and vegetables can counteract oxidative stress, improve health outcomes, and reduce disease risk [5, 8–11, 17]. Growing consumer demand for clean-label and health-promoting foods has led the industry to replace synthetic additives with natural antioxidants from fruits, vegetables, herbs, and spices [18–23].

These natural compounds enhance food quality, extend shelf life, and offer potential health benefits. For example, radish leaf extracts and fruit flours improve antioxidant properties, texture, stability, and nutritional content in products like mayonnaise and meat substitutes [20, 24]. Other plant-derived antioxidants, including rosemary, citrus, oregano, and leafy greens, preserve the color, flavor, and safety of meat products, often matching or surpassing synthetic alternatives while enhancing consumer acceptance [18, 19, 21, 23, 25, 26]. Numerous compounds, such as polyphenols, flavonoids, and vitamins, also exhibit antiinflammatory, antimicrobial, and immune-modulating properties [24, 27–30]. While in vitro and animal studies show promising results, health benefits in humans depend on metabolism and bioavailability, and some clinical trials yield contradictory findings [28, 29].

2. Mechanisms of antioxidant action

Antioxidants use several interconnected mechanisms to protect cells from oxidative damage. The most direct action is scavenging ROS and free radicals by donating electrons or hydrogen atoms, thus neutralizing these harmful molecules and preventing cellular injury [31–33]. Numerous antioxidants, including polyphenols and bioactive peptides, can also activate the Kelch ECH associating protein 1 (Keap1)-nuclear factor erythroid 2–related factor 2 (Nrf2) signaling pathway: this leads to the release and nuclear translocation of the Nrf2 transcription factor, which then upregulates genes encoding antioxidant enzymes like SOD, GPx, and others, boosting the cell's own defense systems [34–38]. Some antioxidants can chelate metal ions (like Fe^{2+}), reducing their ability to catalyze the formation of additional free radicals [33, 34]. Others, such as caffeine and its metabolites, participate in a cascade effect, not only scavenging radicals but also repairing oxidized biomolecules and regenerating other antioxidants [32]. Recent research has revealed novel mechanisms, such as the oxidation of hydrogen sulfide to polysulfides by polyphenols and anthocyanins, which generates secondary cytoprotective molecules and triggers further protective cellular responses [39, 40]. Additionally, the synergistic action of different antioxidants, for example, the combined effects of quercetin and catechin, can regulate microRNAs and inhibit negative regulators like BACH1, further amplifying antioxidant protection [38]. In summary, antioxidant action involves a combination of direct chemical neutralization of ROS, activation of endogenous defense pathways, metal ion chelation, repair of damaged molecules, and modulation of cellular signaling networks [31–40].

3. Major classes of plant-derived antioxidants and their natural sources

Plants produce a wide range of secondary metabolites that are not directly involved in plants' growth, development, or reproduction. However, they are essential for plant defense, ecological balance, and human health (**Table 1**). Many secondary metabolites, such as flavonoids, alkaloids, and terpenoids, show strong antimicrobial and antiviral activities, including potential action against pathogens like SARS-CoV-2 and bacteria. For instance, molecular docking studies of 4,704 plant secondary metabolites revealed that triterpenoids, flavonoids, anthocyanidins, and flavonol glycosides strongly interact with key SARS-CoV-2 proteins involved in viral entry and replication, indicating potential antiviral roles for these compounds [41]. Another study found that flavonoids such as dalpanitin and vicenin-3, isolated from *Derris scandens*, showed notable antimicrobial activity against *Staphylococcus aureus*, *E. coli*, and *P. aeruginosa*, with minimum inhibitory concentrations (MICs) as low as 23 μg/mL, further validating the therapeutic potential of plant-derived secondary metabolites [42].

Class of metabolite	Subtypes	Natural sources	Biological functions	Reference
Phenolics	Flavonoids, phenolic acids, tannins	Fruits, vegetables, and medicinal plants	Antioxidant, antimicrobial, antiinflammatory	[44, 46–51]
Alkaloids	Lycorine, ungeremine, ambelline	Medicinal plants (e.g., *Crinum glaucum*)	Antimicrobial, therapeutic	[44, 45]
Terpenoids	Triterpenoid, monoterpenes, sesquiterpenes	Herbs, roots, leaves	Antiviral, antiinflammatory, aromatic properties	[41, 44, 49]
Coumarins	Daphnetin	Various plants	Plant defense against pathogens; antimicrobial	[43]
Glucosinolates	Various glucosinolate derivatives	*Capparis spinosa*, Brassicaceae family	defense mechanisms, health benefits	[44, 52]
Quinones	Alkannin, shikonin, asperpentyn	Roots of Boraginaceae, endophytes	antimicrobial, wound healing	[53, 54]
Steroids and saponins	β-sitosterol, saponins	Medicinal plants	antimicrobial, health-promoting activities	[44, 45]

Table 1.
Major classes of plant secondary metabolites, their examples, natural sources, and biological functions.

The coumarin compound daphnetin significantly reduced virulence in *Ralstonia solanacearum* by inhibiting extracellular polysaccharide (EPS) production and biofilm formation, indicating its role as a natural antivirulence agent [43]. Similarly, various alkaloids, including ungeremine, lycorine, and others isolated from *Crinum glaucum*, demonstrated antibacterial properties, with ungeremine showing the most potent activity (MIC = 8 µg/mL) against *Enterococcus faecalis* and *Pseudomonas aeruginosa* [44]. Finally, methanol and dichloromethane extracts of *Capparis spinosa* revealed the presence of 21 secondary metabolites (e.g., flavonoids, glucosinolates, alkaloids, triterpenes) that exhibited strong antioxidant and enzyme inhibition activities, supporting its traditional use and potential for novel bioactive drug development [45].

Many of these secondary metabolites also act as antioxidants, helping to neutralize harmful free radicals. Plant-derived antioxidants are a diverse group of compounds that help protect cells from oxidative damage. Phenolic compounds, especially flavonoids, are potent antioxidants found in a wide range of plants. This has been supported by several studies. One study highlighted that fruits, vegetables, seeds, and nuts accumulate phenolic compounds with significant reducing capacity, confirmed through Folin-Ciocalteu and other antioxidant assays, and identified flavonoids such as flavonols and flavanones as key contributors [47]. In another investigation, methanolic extracts of *Capparis spinosa* were found to contain high levels of phenolics and flavonoids, which strongly correlated with radical scavenging activity and other bioactive properties [44]. Research on bamboo revealed the presence of 97 flavonoids and emphasized their dietary significance, illustrating the widespread distribution of these compounds in the plant kingdom [50]. Furthermore, a study using *Solidago nemoralis* hairy root cultures demonstrated that nanoharvested polyphenolic flavonoids exhibited strong antiradical (antioxidant) activity, confirming their biological functionality [46]. Lastly, although primarily focused on anticancer activity, a study on *Origanum vulgare* characterized its extract as rich in plant secondary metabolites and demonstrated its safety and bioactivity, supporting the therapeutic relevance of such compounds [51].

Secondary metabolites are the basis for many traditional medicines and are being explored for drug discovery and therapeutic applications. This is exemplified by studies on ethnopharmacologically important plants, such as those used in Sri Lankan traditional medicine, where bioactivity-guided screening identified potent antimicrobial flavonoids like dalpanitin and vicenin-3 from *Derris scandens*, demonstrating measurable antibacterial effects against pathogens, including *Staphylococcus aureus* and *E. coli* [42]. Likewise, *Capparis spinosa* L. extracts, rich in flavonoids and phenolics, exhibited strong antioxidant properties alongside enzyme inhibition relevant to neurodegenerative and metabolic diseases, validating its traditional use and highlighting its potential in drug design [44]. In *Alkanna tinctoria*, alkannin and shikonin – secondary metabolites with antimicrobial, antitumoral, and wound-healing properties – were shown to be influenced by beneficial root endophytes, suggesting the promise of biotechnological enhancement of therapeutic compound production [54]. Transcriptome analysis of the endangered medicinal plant *Nardostachys jatamansi* revealed high expression of genes involved in phenolic and terpenoid biosynthesis, with environmental factors modulating metabolite levels linked to therapeutic effects, offering sustainable approaches to medicinal compound production [49]. Further, bioactive metabolites from the fungal endophyte *Curvularia sp*., isolated from *Sapindus saponaria* L., demonstrated antioxidant and enzyme inhibitory activities, identifying (-)-asperpentyn as a compound of pharmaceutical interest [53]. Investigations of *Crinum glaucum* leaf extracts revealed potent antibacterial alkaloids like ungeremine with low MIC values against pathogens, scientifically validating its traditional medicinal applications against infections [45]. Lastly, a hydroalcoholic extract of *Origanum vulgare* L. showed selective anticancer activity by inducing apoptosis and necroptosis in melanoma cells without toxicity to healthy cells, underlining the therapeutic potential of plant secondary metabolites in cancer treatment [51]. Collectively, these findings underscore the critical role of plant and microbe-derived secondary metabolites in traditional medicine and their expanding utility in modern pharmacology and drug discovery. The production of secondary metabolites is often regulated by environmental factors, plant-microbe interactions, and specific genetic pathways, such as those involving transcription factors in *Salvia miltiorrhiza* or endophytic bacteria in *Alkanna tinctoria* [48, 49, 54].

4. Major classes of plant-derived antioxidants: Natural sources, functional roles, and food industry applications

Plant-derived antioxidants, including polyphenols, flavonoids, phenolic acids, carotenoids, and other secondary metabolites, are among the most extensively studied natural compounds due to their proven ability to stabilize foods, inhibit oxidative reactions, and enhance overall shelf life [55, 56]. Their antioxidant activity is primarily governed by hydrogen atom transfer (HAT) and single electron transfer (SET) mechanisms, through which these compounds neutralize free radicals and ROS, thereby interrupting oxidative chain reactions in biological and food systems [55, 56]. Phenolic compounds form the largest and most diverse category of such antioxidants and include subclasses such as flavonols, flavones, anthocyanins, biflavonoids, hydroxybenzoic acids, and hydroxycinnamic acids – each contributing distinct radical-scavenging properties [55, 56]. Numerous studies have demonstrated strong correlations between total phenolic content (TPC), total flavonoid content (TFC), and antioxidant capacity assessed using widely applied assays such as DPPH, ABTS, FRAP, oxygen radical absorbance capacity (ORAC), and CUPRAC, which rely on HAT and/or SET reaction principles [55, 56]. Similar approaches using DPPH, ABTS, iron-reduction capacity, hydroxyl radical scavenging, and LDL oxidation inhibition assays have been applied to Lamiaceae herb extracts (oregano, rosemary, sage, and thyme), showing assay-dependent variations in antioxidant activity and highlighting the important role of specific phenolics, such as rosmarinic acid, in radical scavenging [57]. The consistency of these findings across plant extracts highlights the central role of phenolics and related metabolites in antioxidant defense, while also suggesting that the activity may not always be fully predicted by TPC alone [55–57].

For instance, *Ficus carica* latex from different cultivars exhibited high TPC that corresponded directly with increased DPPH, ABTS, and FRAP values, confirming the key contribution of phenolics to antioxidant capacity [58]. This trend is consistent with results from *Taraxacum officinale*, *Eryngium foetidum*, *Trifolium pratense*, *T. repens*, *Ensete superbum*, multiple wild edible species, and Lamiaceae herbs such as oregano, rosemary, sage, and thyme, whose extracts showed strong antioxidant activities as determined by validated analytical methods, including Folin–Ciocalteu, FRAP, DPPH, ABTS, UPLC-ESI-MS/MS, and LC–MS/MS [59–62]. Similarly, *Astragalus armatus* extract, rich in rosmarinic and chlorogenic acids, demonstrated strong antioxidant activity across multiple assays (CUPRAC, reducing power, beta-carotene, hydroxyl radical) and highlights its potential application in food, cosmetics, and pharmaceutical industries [63]. These findings collectively support the established relationship between phenolic concentration, phenolic composition, and antioxidant efficacy across diverse plant sources.

These antioxidant activities are closely associated with the presence and structural characteristics of hydroxybenzoic and hydroxycinnamic acids, flavonols, anthocyanins, flavones, biflavonoids, and related polyphenolic compounds [64–66]. Their antioxidant capacity is strongly influenced by the number and configuration of hydroxyl groups, which determine their radical-scavenging and reducing abilities [65]. Beyond their chemical properties, these natural antioxidants play a crucial role in biological systems, protecting against oxidative damage and supporting cellular antioxidant defenses [64, 66].

Importantly, plant-derived antioxidants have demonstrated broad applicability as natural food additives and preservatives. *Citrus aurantium* and *C. limon* peel, *Rosa canina* seed, and various herbs (*Sideritis* spp., *Hypericum perforatum*, *Origanum dictamnus*, *Mentha pulegium*, *Melissa officinalis*, *Mentha spicata*, *Lavandula angustifolia*) have been successfully incorporated into dairy products such as yogurt, kefir, cream cheese, and vegan yogurt, maintaining acceptable sensory characteristics while increasing antioxidant capacity [64]. Fermented dandelion extracts have shown enhanced flavonoid content and antioxidant activity, highlighting fermentation as a valuable tool for developing functional food ingredients [59]. Other promising plant sources, including *Eryngium foetidum* [60], *Ficus carica* [58], red and white clover [61], wild banana (*Ensete superbum*) [62], and wild edible species [67], also provide natural antioxidant compounds that can be used for food fortification, preservation, and nutritional enhancement. Additionally,

Antioxidant class/subtype	Plant source/ species	Observed antioxidant/ biological activity	Method/assay	Food application	References
Polyphenols (flavonoids, phenolic acids)	*Citrus aurantium* peel, *C. limon* peel, *Rosa canina* seed; herbs: *Sideritis spp.*, *Hypericum perforatum*, *Origanum dictamnus*, *Mentha pulegium*, *Melissa officinalis*, *Mentha spicata*, *Lavandula angustifolia*	Strong antioxidant activity; bioavailability 4–68%	Folin–Ciocalteu, FRAP, mass spectrometry	Fortification of dairy products (yogurt, kefir, cream cheese, vegan yogurt); improved sensory acceptability (~3/5)	[64]
Flavonoids	*Taraxacum officinale* (dandelion), fermented	Enhanced flavonoid content (183.72 mg/g) and antioxidant activity compared to unfermented	UPLC-ESI-MS/MS, Response Surface Methodology	Functional food additive: potential for fermented foods	[59]
Phenolic compounds (chlorgenic acid, others)	*Eryngium foetidum* leaves	ROS and RNS scavenging (IC50 45–1,000 μg/mL); high antioxidant potential	Ultrasound-assisted extraction; in vitro ROS/RNS assays	Natural antioxidant additive for foods and nutraceuticals	[60]
Polyphenols	*Ficus carica* latex (18 cultivars)	Antioxidant activity (DPPH, ABTS, FRAP): highest in "White Genoa"	Maceration, ultrasound-assisted extraction (UAE), Folin–Ciocalteu	Food supplements, natural additives, drug synthesis potential	[58]
Polyphenols	Red clover (*Trifolium pratense*), White clover (*T. repens*)	High antioxidant potential; increased antioxidant enzyme activity in cell cultures	DPPH, FRAP, Spectrophotometry	Functional food supplement; improves oxidative stress response in cell cultures	[61]
Polyphenols, anthocyanins, flavonols, flavones, biflavonoids, phenolic acids	*Ensete superbum* (wild banana) peel, seed, flower, bract	DPPH IC50 21.97 μg/mL; ORAC: 207.97 μM TE/g; highest cellular antioxidant in bract aqueous extract	RP-HPLC, LC/MS-ESI-TOF; in vitro and ex vivo models	Bract aqueous extract is most effective; it has potential as a functional food ingredient and for pharmaceutical use.	[62]

Antioxidant class/subtype	Plant source/ species	Observed antioxidant/ biological activity	Method/assay	Food application	References
Polyphenols, flavonoids	28 wild edible plants from the Loktak Lake wetland ecosystem	Antioxidant capacity: 1.71–263.7 μM TEAC/g; highest in *Jussiaea repens*, *Gynura cusimbua*.	DPPH, ABTS, FRAP assays	Source of natural antioxidant additives, nutritional supplements, and functional foods	[67]
Polyphenols (rosmarinic, protocatechuic, chlorogenic acids; hyperoside, hesperidin, luteolin, quercetin, naringenin, and kaempferol)	*Astragalus armatus*	Strong antioxidant activity; AChE inhibition: 40.25 ± 1.41 μg/mL; neuroprotective effects	UAE, LC–MS/MS, CUPRAC, beta-carotene, reducing power, hydroxyl radical, molecular docking	Applications in cosmetics, food additives, pharmaceuticals; drug-like polyphenols	[63]

Table 2.
Major classes of natural antioxidants: their examples, sources, antioxidant functions, and food industry applications.

Astragalus armatus, rich in rosmarinic and chlorogenic acids, has demonstrated strong antioxidant and neuroprotective effects, supporting its potential in both food and nutraceutical applications [63].

The practical use of these natural antioxidants improves oxidative stability, extends shelf life, enhances nutritional value, and contributes positively to the sensory quality of food products. Their bioavailability, effectiveness, and compatibility with food matrices make them promising candidates to replace synthetic antioxidants in modern food systems. A comprehensive summary of the antioxidant classes, plant sources, biological activities, analytical methods, and food applications is presented in **Table 2**.

5. Application of plant-derived antioxidants as preservatives and food additives

Natural antioxidants are increasingly used in food preservation to extend shelf life, prevent spoilage, and maintain nutritional quality by inhibiting oxidation and microbial growth. Oxidative degradation of lipids, proteins, and pigments significantly reduces food quality, leading to rancidity, discoloration, and nutrient loss. Synthetic antioxidants, including BHA, BHT, tert-butylhydroquinone (TBHQ), and propyl gallate, are efficient but are increasingly limited due to safety concerns [68].

Plant-derived antioxidants from spices and herbs [69, 70], citrus peels [71], edible flowers [72], fruits [73], and wild plants [74] inhibit lipid oxidation and enhance oxidative stability in food systems. Spices and herbs, rich in phenolic compounds such as flavonoids, terpenoids, lignans, polyphenolics, carotenoids, coumarins, and other bioactive constituents, have long been used to enhance flavor, color, and aroma while effectively preventing oxidative rancidity in foods [75]. Their antioxidant forms, including ground spices/herbs, extracts, emulsions, and encapsulated preparations, help retard lipid oxidation and preserve food quality [75].

Moreover, plant-derived compounds such as essential oils (from *Litsea cubeba*, *clove*, *basil*, and *thyme*), polyphenols (from pomegranate, orange, and banana peels), and polysaccharides (from seaweed like *Porphyra haitanensis*) have demonstrated strong antioxidant and antimicrobial activities, making them effective natural preservatives [70, 76–84]. These substances can be incorporated directly into foods, used in edible films, or encapsulated to improve stability and sensory properties. For example, essential oils and peel extracts have been shown to inhibit the growth of foodborne pathogens and spoilage bacteria, reduce lipid and protein oxidation in meat products, and maintain food quality during storage [70, 77–80, 82–84].
A practical example of plant-based preservatives is provided by a study on basil (*Ocimum basilicum*) essential oil [70]. The authors reported that incorporating 0.2–0.3% basil essential oil into chicken nuggets significantly reduced lipid oxidation (lower thiobarbituric acid reactive substances (TBARS) values) while maintaining good sensory acceptability. Likewise, clove essential oil extracted from *Syzygium aromaticum* buds, dominated by eugenol (66%), exhibited significant antioxidant and antimicrobial activity when incorporated into millet starch edible films, where increasing concentrations from 1–3% (w/w) enhanced both antioxidant capacity and microbial inhibition, with the 3% formulation showing the strongest effects [79]. Similarly, *Litsea cubeba* essential oil demonstrated strong practical preservative potential, with vapor-phase applications inhibiting microbial growth on apples, pears, potatoes, and kohlrabi, and concentrations of 1.46–3.52 μL/mL producing potent antimicrobial effects, while 500 μL/mL suppressed over 50% of tested bacteria and yeasts [80].

Such findings support the growing shift toward natural antioxidants, which not only address consumer concerns about the safety of synthetic preservatives but also allow the valorization of food industry by-products, such as fruit peels [78]. In this context, pomegranate peel extract, rich in punicalagin, has shown strong antioxidant and lipid-protective activity in multiple assays, indicating its potential use as a natural preservative to reduce lipid oxidation in food products [76].

Building on their functional roles in food preservation and safety, several plant-derived antioxidants have also been authorized as food additives, with regulated limits for use in various food products. For instance, yellow pigment extracted from turmeric (*Curcuma longa*) rhizomes demonstrates strong antioxidant and antimicrobial activities, with oil-soluble pigment at 0.2% (w/w) improving oxidative stability in soybean oil, and water-soluble pigment at 3% (w/w) enhancing color, flavor, and sensory attributes in chicken breast fillets while reducing microbial load and extending shelf-life up to 39 days when combined with gamma irradiation [85]. Curcumin (E100), another natural colorant, has been shown to confer color while exhibiting bioactivity, and different formulations such as water-dispersible and nanoencapsulated curcumin improve its stability and applicability in hydrophilic foods like yogurt [85]. Similarly, *Indigofera tinctoria* leaf extract (ITLE), encapsulated with maltodextrin, has been successfully incorporated into ice cream at levels of 0–1.8%, with 1.2% providing optimal sensory acceptance [86]. The addition of ITLE increased TPC and antioxidant activity, improved color parameters, and influenced textural properties, demonstrating its practical application as a natural antioxidant and colorant in food products. In addition, natural indigo dyes derived from *Indigofera tinctoria* have been evaluated for food applications [87]. Analyses using UV-Vis spectrophotometry and FTIR showed stable color properties at concentrations of 0.002–0.008 mg/mL, with maximum absorption at 550–700 nm and strong staining ability at pH ~ 9. FTIR spectra confirmed characteristic functional groups, indicating that indigo can be used as a stable natural blue dye in foods and beverages, potentially reducing reliance on synthetic dyes [87].

In addition to natural indigo dyes, chamomile (*Matricaria recutita* L.) essential oil has been evaluated as a natural food additive [88]. When incorporated into cakes at 0.05–0.15%, chamomile essential oil demonstrated significant antioxidant and antimicrobial activity, reducing lipid oxidation and inhibiting yeast and mold growth over 75 days of storage [88]. While its activity was slightly lower than that of synthetic antioxidants and preservatives (TBHQ and potassium sorbate), chamomile essential oil maintained sensory quality and offered a safe, natural alternative without adverse health effects [88]. Complementary studies using chamomile flower powder in broiler diets (1 kg per 100 kg feed) reported improved growth performance, enhanced feed efficiency, and reduced mortality, highlighting its bioactive richness and safety [89]. Together, these findings indicate that chamomile – whether as essential oil or flower powder – possesses functional antioxidant and antimicrobial properties that support its application as a natural food additive or preservative.

Further supporting the role of plant-based additives, steviol glycosides from *Stevia rebaudiana* – a calorie-free natural sweetener rich in bioactive compounds – have undergone extensive safety evaluation by the Joint FAO/WHO Expert Committee on Food Additives (JECFA) and the Scientific Committee on Food (SCF) [90]. An acceptable daily intake (ADI) of 4 mg/kg body weight/day (expressed as steviol equivalents) has been established, with toxicological assessments demonstrating no carcinogenic, genotoxic, or reproductive/developmental risks within this limit [90]. Although steviol glycosides are widely used as natural sweeteners and functional ingredients, exposure assessments indicate that the ADI may be exceeded at maximum proposed use levels, emphasizing the need for controlled formulation in food applications [90].

Overall, these findings demonstrate that many plant-derived compounds – ranging from pigments and essential oils to steviol glycosides – are not only effective natural antioxidants and colorants but also meet established safety standards, supporting their broader use as regulated, clean-label food additives.

6. Natural antioxidant extraction methods, their effects on function and stability

Natural antioxidants are extracted from plants and other natural sources using a variety of methods, each of which affects yield, purity, and stability. Extraction parameters – such as temperature, solvent type, extraction time, and liquid-to-solid ratio – are critical for optimizing both yield and antioxidant potency [91–97].

Modern extraction techniques, such as ultrasound-assisted extraction (UAE), microwave-assisted extraction (MAE), pressurized-liquid extraction (PLE), and supercritical or subcritical CO_2 extraction, are favored for their efficiency, sustainability, and ability to maximize antioxidant content compared to conventional solvent extraction [91–93, 98–104]. The use of natural deep eutectic solvents (NADES) has emerged as a green alternative, improving polyphenol yield and antioxidant activity while reducing environmental impact [93, 94, 98, 105–109].

Practical studies have shown that different extraction techniques applied to black pepper, *Withania somnifera*, or seaweed produce different outcomes for extract yield, phytochemical content, and antioxidant activity, demonstrating that "best yield" and "best activity" do not always correspond to the same extract [96, 97, 99]. Predictive modeling approaches, such as response surface methodology (RSM), enable optimization of extraction conditions, standardization of protocols, and industrial-scale applications [99]. Advanced techniques, such as pressurized hot water extraction (PHWE), subcritical water, ionic liquids, and electrohydrodynamic methods, show promise for efficient recovery of antioxidants and thermosensitive compounds, although they may face challenges in scale-up and cost [100]. Green extraction technologies, including ultrasound- and microwave-assisted methods, improve efficiency, reduce solvent use, and enhance sustainability, bridging laboratory-scale research to industrial applications [101].

Regarding stability, phenolic compounds and anthocyanins can degrade under high temperatures, prolonged storage, or extreme pH, although some antioxidant activity may remain stable after moderate heat treatment [102]. Strategies such as encapsulation, edible films, and careful control of processing conditions can help maintain the stability of natural antioxidants in food, supplement, and cosmetic applications [98, 103, 110]. Collectively, the choice of extraction method, parameter optimization, and postextraction handling are critical to obtaining high-quality, stable, and sustainable natural antioxidants suitable for commercial and industrial use.

7. Health benefits and functional role

Natural antioxidants, found in a wide range of foods such as fruits, vegetables, mushrooms, and even certain bacteria, play vital roles in promoting health and preventing disease. Their primary functional role is to neutralize harmful free radicals and reduce oxidative stress, which is linked to aging, neurodegenerative diseases, diabetes, and chronic inflammation [66, 103–104, 110–116].

In addition to antioxidant activity, several plant-derived natural products exhibit antimicrobial and antiviral properties, contributing to host defense against pathogenic microorganisms and viral infections, as demonstrated in multiple in-vitro and in-vivo studies. For example, compounds like caffeoylquinic acids, astaxanthin, and polyphenols can protect cells from lipid peroxidation, inhibit protein glycation, and support the body's own antioxidant enzyme systems, such as SOD and CAT [66, 103, 104, 111, 113].

Similarly, in vitro studies have demonstrated antimicrobial and antigenotoxic properties of certain plant products: essential oils and phenolic constituents of *Achillea millefolium* (yarrow) showed antimicrobial activity against *Candida albicans*, *Staphylococcus epidermidis*, *Escherichia coli*, and *Klebsiella pneumoniae*, although with relatively high minimal inhibitory concentration (MIC) values (>1.5 mg mL^{-1}), indicating modest potency [117]. Additionally, *Baccharis trimera* extracts exhibited antioxidant and antigenotoxic effects in cellular assays [118].

Natural antioxidants also show promise in supporting cognitive function and protecting against neurodegenerative diseases like Alzheimer's by reducing inflammation, modulating signaling pathways (such as Nrf2 and Janus kinase 2/signal transducer and activator of transcription 3 [JAK2/STAT3]), and preventing the buildup of toxic proteins [112, 114, 116]. Beyond neuroprotection, these compounds frequently exert immunomodulatory and antiinflammatory effects by regulating cytokine production and oxidative signaling pathways, reinforcing their systemic protective roles. Some, like polysaccharides from *Dendrobium officinale*, not only enhance antioxidant defenses but also boost immune function [111].

Limited human clinical trials, such as a randomized study on *licorice* pastille, indicate symptom improvement and good tolerability, supporting potential complementary use in humans [119]. In vivo evidence from animal models further supports systemic benefits: *Baccharis trimera* extracts reduced oxidative stress, improved liver and cardiovascular function, and mitigated damage in hypertensive and dyslipidemic rats [120], while *Bauhinia forficata* leaf preparations relaxed rat aortic rings via nitric oxide and potassium channel–dependent pathways, demonstrating cardiovascular benefits [121].

Consistent with this multifunctionality, *Nigella sativa* (black seed) has been reported to possess antioxidant, antiinflammatory, immunomodulatory, and antiviral activities, with potential benefits in managing viral infections such as Langya henipavirus through supportive and symptom-targeted mechanisms [122].

However, safety and efficacy are dose- and context-dependent, as animal studies have shown that certain extracts may exert both protective and adverse effects depending on the concentration and biological target.

Additionally, natural antioxidants can help prevent complications of diabetes by inhibiting the formation of advanced glycation end-products and protein aggregation [66]. Their safety, availability, and multifunctional benefits make them attractive for use in functional foods, supplements, and even targeted therapies for specific organs, such as the liver [104, 115].

These broad pharmacological properties highlight their potential as complementary agents alongside conventional therapies, particularly where treatment options remain largely supportive. Overall, natural antioxidants contribute to health maintenance, disease prevention, and improved quality of life through their diverse biological activities.

8. Regulatory status and labeling

Common antioxidants such as ascorbic acid (vitamin C), tocopherols (vitamin E), BHA, BHT, TBHQ, and gallates are widely used in food and feed, and many have achieved generally recognized as safe (GRAS) status in the USA, allowing their use without premarket approval when used within established limits [123]. In the European Union (EU), the European Food Safety Authority (EFSA) authorizes antioxidants as food or feed additives, assigning E-numbers and setting maximum permitted levels, but the regulatory process can be more restrictive, especially regarding claims about in vivo efficacy and health benefits [124]. Regulatory approval and permitted uses can differ between regions, so an antioxidant approved in the USA may not automatically be allowed in the EU or other markets [123, 124]. For labeling, both the Food and Drug Administration (FDA) and EFSA require that antioxidant claims be truthful and not misleading; products can only make "antioxidant" claims if they contain recognized antioxidant nutrients at significant levels and if the claim is supported by scientific evidence [123, 124]. Labels must typically list the antioxidant by its common or chemical name and, in some cases, its function (such as "to preserve freshness"). Health claims related to disease prevention or treatment are more strictly regulated and generally require preapproval based on substantial scientific substantiation [124]. The regulatory landscape is designed to ensure that antioxidants used in foods and feeds are safe, properly labeled, and that any health-related claims are evidence-based and not misleading to consumers [123, 124].

9. Challenges and future trends

Ensuring the stability and standardization of natural antioxidant extracts is a persistent challenge, as antioxidant activity can vary significantly depending on extraction methods, plant species, and the specific assay used, making it difficult to guarantee consistent quality and compare results across studies [125–127]. The cost and availability of plant materials further complicate large-scale production, especially as demand for sustainable and high-quality sources rises. Biotechnology and fermentation are emerging as promising trends, enabling the production of antioxidant compounds through engineered microbes or optimized fermentation processes, which can reduce reliance on agricultural resources and improve yield consistency. Nanotechnology is also advancing the field, with nanoformulations and nanozymes (nanomaterials with enzyme-like antioxidant activity) being developed to enhance delivery, bioavailability, and functionality of antioxidants in food and agricultural applications [128]. However, the use of nanomaterials introduces safety and regulatory concerns, as their effects can shift from beneficial to harmful depending on dose and exposure route, and excessive concentrations may even induce oxidative stress rather than prevent it [128]. Future directions will likely focus on more precise extraction and quantification methods, improved standardization, and innovative production and delivery systems, while balancing safety, cost, and sustainability [125–128].

10. Conclusion

Plant-derived antioxidants are vital bioactive compounds with significant applications in both food preservation and human health. Their ability to neutralize ROS, chelate metal ions, and modulate cellular defense mechanisms underpins their protective roles against oxidative damage. As natural alternatives to synthetic preservatives, these compounds enhance food shelf life, maintain nutritional quality, and meet consumer demand for clean-label products. Their use spans dairy products, functional foods, beverages, and supplements, providing safe and functional alternatives to synthetic additives. Advances in extraction technologies, encapsulation, and sensory optimization improve the stability, bioavailability, and acceptability of these antioxidants, supporting their integration into a wide range of foods. However, challenges remain, including variability in antioxidant content due to plant sources and extraction methods, regulatory hurdles, and the need for standardized assays to accurately measure antioxidant capacity. Emerging approaches, such as biotechnology-driven production and nanotechnology-based delivery systems, offer promising solutions. Continued interdisciplinary research and harmonized regulatory frameworks will be essential to fully realize the potential of plant-derived antioxidants in sustainable food systems, functional foods, and health promotion.

Acknowledgments

The authors acknowledge the use of ChatGPT for language polishing and editing assistance during the preparation of this manuscript.

Conflict of Interest

The authors declare no conflict of interest.

Author details

Nawab Nasir[1]*, Din Mohammad Mushfiq[1] and Sevinç Ateş[2]

1 Department of Horticulture, Faculty of Plant Sciences, Afghanistan National Agricultural Sciences and Technology University (ANASTU), Kandahar, Afghanistan

2 Department of Horticulture, Faculty of Agriculture, Akdeniz University, Antalya, Türkiye

*Address all correspondence to: nasir.redtomato@gmail.com

References

[1] Valenzuela A, Sanhueza J, Nieto S. Natural antioxidants in functional foods: From food safety to health benefits. Grasas Y Aceites. 2003;**54**:295–303. DOI: 10.3989/gya.2003.v54.i3.245

[2] Jomova K, Alomar SY, Alwasel SH, Nepovimova E, Kuca K, Valko M. Several lines of antioxidant defense against oxidative stress: Antioxidant enzymes, nanomaterials with multiple enzyme-mimicking activities, and low-molecular-weight antioxidants. Archives of Toxicology. 2024;**98**:1323–1367. DOI: 10.1007/s00204-024-03696-4

[3] Ali SS, Ahsan H, Zia MK, Siddiqui T, Khan FH. Understanding oxidants and antioxidants: Classical team with new players. Journal of Food Biochemistry. 2020;**44**:e13145. DOI: 10.1111/jfbc.13145

[4] Sies H. Oxidative stress: Concept and some practical aspects. Antioxidants. 2020;**9**:e852. DOI: 10.3390/antiox9090852

[5] Jiang S, Liu H, Li C. Dietary regulation of oxidative stress in chronic metabolic diseases. Foods. 2021;**10**:1854. DOI: 10.3390/foods10081854

[6] Pizzino G, Irrera N, Cucinotta M, Pallio G, Mannino F, Arcoraci V, Squadrito F, Altavilla D, Bitto A. Oxidative stress: Harms and benefits for human health. Oxidative Medicine and Cellular Longevity. 2017;**2017**:8416763. DOI: 10.1155/2017/8416763

[7] Ponnampalam EN, Kiani A, Santhiravel S, Holman BWB, Lauridsen C, Dunshea FR. The importance of dietary antioxidants on oxidative stress, meat and milk production, and their preservative aspects in farm animals: Antioxidant action, animal health, and product quality—invited review. Animals. 2022;**12**:3279. DOI: 10.3390/ani12233279

[8] Ávila-Escalante ML, Coop-Gamas F, Cervantes-Rodríguez M, Méndez-Iturbide D, Aranda-González II. The effect of diet on oxidative stress and metabolic diseases—Clinically controlled trials. Journal of Food Biochemistry. 2020;44e13191. DOI: 10.1111/jfbc.13191

[9] Vetrani C, Costabile G, Di Marino L, Rivellese AA. Nutrition and oxidative stress: A systematic review of human studies. International Journal of Food Sciences and Nutrition. 2013;**64**:312–326. DOI: 10.3109/09637486.2012.738651

[10] Wang Y, Chen Y, Zhang X, Lu Y, Chen H. New insights in intestinal oxidative stress damage and the health intervention effects of nutrients: A review. Journal of Functional Foods. 2020;**75**:104248. DOI: 10.1016/j.jff.2020.104248

[11] Sharifi-Rad M, Anil Kumar NV, Zucca P, Varoni EM, Dini L, Panzarini E, Rajkovic J, Tsouh Fokou PV, Azzini E, Peluso I, et al. Lifestyle, oxidative stress, and antioxidants: Back and forth in the pathophysiology of chronic diseases. Frontiers in Physiology. 2020;**11**:552535

[12] Reddy VP. Oxidative stress in health and disease. Biomedicines. 2023;**11**:2925. DOI: 10.3390/biomedicines11112925

[13] Chib A, Gupta N, Bhat A, Anjum N, Yadav G. Role of antioxidants in food. International Journal of Chemical Studies. 2020;**8**:2354–2361. DOI: 10.22271/chemi.2020.v8.i1aj.8621

[14] Gulcin İ. Antioxidants and antioxidant methods: An updated overview. Archives of Toxicology. 2020;**94**:651–715. DOI: 10.1007/s00204-020-02689-3

[15] Gülçin İ. Antioxidant activity of food constituents: An overview. Archives of Toxicology. 2012;**86**:345–391. DOI: 10.1007/s00204-011-0774-2

[16] Oroian M, Escriche I. Antioxidants: Characterization, natural sources, extraction and analysis. Food Research International. 2015;**74**:10–36. DOI: 10.1016/j.foodres.2015.04.018

[17] Battino M, Giampieri F, Cianciosi D, Ansary J, Chen X, Zhang D, Gil E, Forbes-Hernández T. The roles of strawberry and honey phytochemicals on human health: A possible clue on the molecular mechanisms involved in the prevention of oxidative stress and inflammation. Phytomedicine. 2021;**86**:153170. DOI: 10.1016/j.phymed.2020.153170

[18] Martínez L, Bastida P, Castillo J, Ros G, Nieto G. Green alternatives to synthetic antioxidants, antimicrobials, nitrates, and nitrites in clean label Spanish Chorizo. Antioxidants. 2019;**8**:184. DOI: 10.3390/antiox8060184

[19] Martínez-Zamora L, Peñalver R, Ros G, Nieto G. Substitution of synthetic nitrates and antioxidants by spices, fruits and vegetables in clean label Spanish chorizo. Food Research International. 2021;**139**:109835. DOI: 10.1016/j.foodres.2020.109835

[20] Vieira MR, Simões S, Carrera-Sánchez C, Raymundo A. Development of a clean label mayonnaise using fruit flour. Foods. 2023;**12**:2111. DOI: 10.3390/foods12112111

[21] Ribeiro JS, Santos MJMC, Silva LKR, Pereira LCL, Santos IA, da Silva Lannes SC, da Silva MV. Natural antioxidants used in meat products: A brief review. Meat Science. 2019;**148**:181–188. DOI: 10.1016/j.meatsci.2018.10.016

[22] Nieto G, Martínez-Zamora L, Peñalver R, Marín-Iniesta F, Taboada-Rodríguez A, López-Gómez A, Martínez-Hernández GB. Applications of plant bioactive compounds as replacers of synthetic additives in the food industry. Foods. 2024;**13**:47. DOI: 10.3390/foods13010047

[23] Oswell NJ, Thippareddi H, Pegg RB. Practical use of natural antioxidants in meat products in the U.S.: A review. Meat Science. 2018;**145**:469–479. DOI: 10.1016/j.meatsci.2018.07.020

[24] Goyeneche R, Rodrigues CR, Quispe-Fuentes I, Pellegrini MC, Cumino A, Di Scala K. Radish leaves extracts as functional ingredients: Evaluation of bioactive compounds and health-promoting capacities. Waste and Biomass Valorization. 2025;**16**:1003–1014. DOI: 10.1007/s12649-024-02760-5

[25] Martínez-zamora L, Ros G, Nieto G. Synthetic vs. Natural hydroxytyrosol for clean label lamb burgers. Antioxidants. 2020;**9**:1–15. DOI: 10.3390/antiox9090851

[26] Jiang J, Xiong YL. Natural antioxidants as food and feed additives to promote health benefits and quality of meat products: A review. Meat

Science. 2016;**120**:107–117. DOI: 10.1016/j.meatsci.2016.04.005

[27] Rahaman MM, Hossain R, Herrera-Bravo J, Islam MT, Atolani O, Adeyemi OS, Owolodun OA, Kambizi L, Daştan SD, Calina D, et al. Natural antioxidants from some fruits, seeds, foods, natural products, and associated health benefits: An update. Food Science & Nutrition. 2023;**11**:1657–1670

[28] Zhang X, Gao F. Natural antioxidants for health promotion and disease prevention. Frontiers in Pharmacology. 2014;**5**:266. DOI: 10.3389/fphar.2014.00266

[29] Bjørklund G, Shanaida M, Lysiuk R, Butnariu M, Peana M, Sarac I, Strus O, Smetanina K, Chirumbolo S. Natural compounds and products from an anti-aging perspective. Molecules. 2022;**27**:7084. DOI: 10.3390/molecules27207084

[30] Jideani AIO, Silungwe H, Takalani T, Omolola AO, Udeh HO, Anyasi TA. Antioxidant-rich natural fruit and vegetable products and human health. International Journal of Food Properties. 2021;**24**:41–67. DOI: 10.1080/10942912.2020.1866597

[31] Wang Y-Z, Wang Y-M, Pan X, Chi C-F, Wang B. Antioxidant mechanisms of the oligopeptides (FWKVV and FMPLH) from muscle hydrolysate of miiuy croaker against oxidative damage of HUVECs. Oxidative Medicine and Cellular Longevity. 2021;**2021**. DOI: 10.1155/2021/9987844

[32] Vieira AJSC, Gaspar EM, Santos PMP. Mechanisms of potential antioxidant activity of caffeine. Radiation Physics and Chemistry. 2020;**174**:108968. DOI: 10.1016/j.radphyschem.2020.108968

[33] Wu R, Huang J, Huan R, Chen L, Yi C, Liu D, Wang M, Liu C, He H. New insights into the structure-activity relationships of antioxidative peptide PMRGGGGYHY. Food Chemistry. 2021;**337**:127678. DOI: 10.1016/j.foodchem.2020.127678

[34] Ren L, Fan J, Yang Y, Liu X, Wang B, Bian X, Wang D, Xu Y, Liu B, Zhu P, et al Identification, in silico selection, and mechanism study of novel antioxidant peptides derived from the rice bran protein hydrolysates. Food Chemistry. 2023;**408**:135230

[35] Tonolo F, Fiorese F, Moretto L, Folda A, Scalcon V, Grinzato A, Ferro S, Arrigoni G, Bindoli A, Feller E, et al. Identification of new peptides from fermented milk showing antioxidant properties: Mechanism of action. Antioxidants. 2020;**9**:117

[36] Zhang J, Li M, Zhang G, Tian Y, Kong F, Xiong S, Zhao S, Jia D, Manyande A, Du H. Identification of novel antioxidant peptides from snakehead (Channa argus) soup generated during gastrointestinal digestion and insights into the anti-oxidation mechanisms. Food Chemistry. 2021;**337**:127921. DOI: 10.1016/j.foodchem.2020.127921

[37] Tonolo F, Folda A, Cesaro L, Scalcon V, Marin O, Ferro S, Bindoli A, Rigobello MP. Milk-derived bioactive peptides exhibit antioxidant activity through the Keap1-Nrf2 signaling pathway. Journal of Functional Foods. 2020;**64**:103696. DOI: 10.1016/j.jff.2019.103696

[38] Wang D, Jiang Y, Sun-Waterhouse D, Zhai H, Guan H, Rong X, Li F, Yu J,

Li D. MicroRNA-based regulatory mechanisms underlying the synergistic antioxidant action of quercetin and catechin in H2O2-stimulated HepG2 cells: Roles of BACH1 in Nrf2-dependent pathways. Free Radical Biology and Medicine. 2020;**153**: 122–131. DOI: 10.1016/j.freeradbiomed. 2020.04.018

[39] Olson KR, Gao Y, Briggs A, Devireddy M, Iovino NA, Licursi M, Skora NC, Whelan J, Villa BP, Straub KD. 'Antioxidant' berries, anthocyanins, resveratrol and rosmarinic acid oxidize hydrogen sulfide to polysulfides and thiosulfate: A novel mechanism underlying their biological actions. Free Radical Biology and Medicine. 2021;**165**:67–78. DOI: 10.1016/j.freeradbiomed.2021.01.035

[40] Olson KR, Briggs A, Devireddy M, Iovino NA, Skora NC, Whelan J, Villa BP, Yuan X, Mannam V, Howard S, et al. Green tea polyphenolic antioxidants oxidize hydrogen sulfide to thiosulfate and polysulfides: A possible new mechanism underpinning their biological action. Redox Biology. 2020;**37**:101731

[41] Puttaswamy H, Gowtham HG, Ojha MD, Yadav A, Choudhir G, Raguraman V, Kongkham B, Selvaraju K, Shareef S, Gehlot P, et al. In silico studies evidenced the role of structurally diverse plant secondary metabolites in reducing SARS-CoV-2 pathogenesis. Scientific Reports. 2020;**10**:20584

[42] Mohotti S, Rajendran S, Muhammad T, Strömstedt AA, Adhikari A, Burman R, de Silva ED, Göransson U, Hettiarachchi CM, Gunasekera S. Screening for bioactive secondary metabolites in Sri Lankan medicinal plants by microfractionation and targeted isolation of antimicrobial flavonoids from Derris scandens. Journal of Ethnopharmacology. 2020;**246**:112158. DOI: 10.1016/ j.jep.2019.112158

[43] Yang L, Wei Z, Li S, Xiao R, Xu Q, Ran Y, Ding W. Plant secondary metabolite, daphnetin reduces extracellular polysaccharides production and virulence factors of Ralstonia solanacearum. Pesticide Biochemistry and Physiology. 2021;**179**:104948. DOI: 10.1016/ j.pestbp.2021.104948

[44] Saleem H, Khurshid U, Sarfraz M, Ahmad I, Alamri A, Anwar S, Alamri AS, Locatelli M, Tartaglia A, Mahomoodally MF, et al. Investigation into the biological properties, secondary metabolites composition, and toxicity of aerial and root parts of Capparis spinosa L.: An important medicinal food plant. Food and Chemical Toxicology. 2021;**155**:112404

[45] Kianfé BY, Kühlborn J, Tchuenguem RT, Tchegnitegni BT, Ponou BK, Groß J, Teponno RB, Dzoyem JP, Opatz T, Tapondjou LA. Antimicrobial secondary metabolites from the medicinal plant Crinum glaucum. A. Chev. (Amaryllidaceae). South African Journal of Botany. 2020;**133**:161–166. DOI: 10.1016/ j.sajb.2020.07.026

[46] Khan MA, Wallace WT, Sambi J, Rogers DT, Littleton JM, Rankin SE, Knutson BL. Nanoharvesting of bioactive materials from living plant cultures using engineered silica nanoparticles. Materials Science and Engineering C. 2020;**106**:110190. DOI: 10.1016/j.msec.2019.110190

[47] Platzer M, Kiese S, Herfellner T, Schweiggert-Weisz U, Eisner P. How

does the phenol structure influence the results of the folin-ciocalteu assay? Antioxidants. 2021;**10**:811. DOI: 10.3390/antiox10050811

[48] Liu S, Wang Y, Shi M, Maoz I, Gao X, Sun M, Yuan T, Li K, Zhou W, Guo X, et al. SmbHLH60 and SmMYC2 antagonistically regulate phenolic acids and anthocyanins biosynthesis in Salvia miltiorrhiza. Journal of Advanced Research. 2022;**42**:205–219

[49] Dhiman N, Kumar A, Kumar D, Bhattacharya A. De novo transcriptome analysis of the critically endangered alpine Himalayan herb Nardostachys jatamansi reveals the biosynthesis pathway genes of tissue-specific secondary metabolites. Scientific Reports. 2020;**10**:17186. DOI: 10.1038/s41598-020-74049-1

[50] Wang L, Huang G, Hou R, Qi D, Wu Q, Nie Y, Zuo Z, Ma R, Zhou W, Ma Y, et al. Multi-omics reveals the positive leverage of plant secondary metabolites on the gut microbiota in a non-model mammal. Microbiome. 2021;**9**:142

[51] Nanni V, Di Marco G, Sacchetti G, Canini A, Gismondi A. Oregano phytocomplex induces programmed cell death in melanoma lines via mitochondria and DNA damage. Foods. 2020;**9**:1486. DOI: 10.3390/foods9101486

[52] Xia X, Wang Q, Gurr GM, Vasseur L, Han S, You M. Gut bacteria mediated adaptation of diamondback moth, Plutella xylostella, to secondary metabolites of host plants. mSystems. 2023;**8**:e00826–23. DOI: 10.1128/msystems.00826-23

[53] Polli AD, Dos S RMA, Garcia A, Polonio JC, Santos CM, Silva AA, Orlandelli RC, Castro JC, Abreu-Filho BA, Cabral MRP, et al. Secondary metabolites of Curvularia sp. G6-32, an endophyte of Sapindus saponaria, with antioxidant and anticholinesterasic properties. Natural Product Research. 2021;**35**:4148–4153

[54] Rat A, Naranjo HD, Krigas N, Grigoriadou K, Maloupa E, Alonso AV, Schneider C, Papageorgiou VP, Assimopoulou AN, Tsafantakis N, et al. Endophytic bacteria from the roots of the medicinal plant Alkanna tinctoria Tausch (Boraginaceae): Exploration of plant growth promoting properties and potential role in the production of plant secondary metabolites. Frontiers in Microbiology. 2021;**12**:633488

[55] Pisoschi AM, Pop A, Cimpeanu C, Predoi G. Antioxidant capacity determination in plants and plant-derived products: A review. Oxidative Medicine and Cellular Longevity. 2016;**2016**:9130976. DOI: 10.1155/2016/9130976

[56] Lourenço SC, Moldão-Martins M, Alves VD. Antioxidants of natural plant origins: From sources to food industry applications. Molecules. 2019;**24**:4132. DOI: 10.3390/molecules24224132

[57] Dorman HJD, Peltoketo A, Hiltunen R, Tikkanen MJ. Characterisation of the antioxidant properties of de-odourised aqueous extracts from selected Lamiaceae herbs. Food Chemistry. 2003;**83**:255–262. DOI: 10.1016/S0308-8146(03)00088-8

[58] Shahinuzzaman M, Yaakob Z, Anuar FH, Akhtar P, Kadir NHA, Hasan AKM, Sobayel K, Nour M, Sindi H, Amin N, et al. In vitro antioxidant activity of Ficus carica L. latex from 18 different cultivars. Scientific Reports. 2020;**10**:10852

[59] Liu N, Song M, Wang N, Wang Y, Wang R, An X, Qi J. The effects of solid-state fermentation on the content, composition and in vitro antioxidant activity of flavonoids from dandelion. PLoS One. 2020;**15**:e0239076. DOI: 10.1371/journal.pone.0239076

[60] Do Stc LD, Barbosa-Carvalho APP, de Siqueira FC, E Sousa RP, Lopes AS, Chisté RC. Extracts of Eryngium foetidum Leaves from the Amazonia Were Efficient Scavengers of ROS and RNS. Antioxidants. 2023;**12**:1112. DOI: 10.3390/antiox12051112

[61] Jakubczyk K, Łukomska A, Gutowska I, Kochman J, Janił J, Janda K. Edible flowers extracts as a source of bioactive compounds with antioxidant properties—in vitro studies. Applied Sciences. 2021;**11**:2120. DOI: 10.3390/app11052120

[62] Mathew NS, Negi PS. Phenolic content and anti-oxidative attributes of various parts of wild banana (Ensete superbum Roxb. Cheesman) plant. Journal of Food Biochemistry. 2021; **45**:e13657. DOI: 10.1111/jfbc.13657

[63] Lekmine S, Bendjedid S, Benslama O, Martín-García AI, Boussekine S, Kadi K, Akkal S, Nieto G, Sami R, Al-Mushhin AAM, et al. Ultrasound-Assisted Extraction, LC–MS/MS Analysis, Anticholinesterase, and Antioxidant Activities of Valuable Natural Metabolites from Astragalus armatus Willd. In Silico Molecular Docking and In Vitro Enzymatic Studies. Antioxidants. 2022;**11**:2000

[64] Kandyliari A, Potsaki P, Bousdouni P, Kaloteraki C, Christofilea M, Almpounioti K, Moutsou A, Fasoulis CK, Polychronis LV, Gkalpinos VK, et al. Development of Dairy Products Fortified with Plant Extracts: Antioxidant and Phenolic Content Characterization. Antioxidants. 2023;**12**:500

[65] Liu J, Du C, Beaman HT, Monroe MBB. Characterization of phenolic acid antimicrobial and antioxidant structure–property relationships. Pharmaceutics. 2020;**12**:419. DOI: 10.3390/pharmaceutics12050419

[66] Liu H, Huo X, Wang S, Yin Z. The inhibitory effects of natural antioxidants on protein glycation as well as aggregation induced by methylglyoxal and underlying mechanisms. Colloids and Surfaces B: Biointerfaces. 2022;**212**:112360. DOI: 10.1016/j.colsurfb.2022.112360

[67] Singh TS, Roy SS, Kshetri P, Ansari MA, Sharma SK, Verma MR, Singh IM, Prakash N, Kandpal B. Comparative study on phenolic, flavonoids and in vitro antioxidant activity of wild edible plants from Loktak Lake wetland ecosystem under North East Indian Himalayan Region. Natural Product Research. 2021;**35**:6045–6048. DOI: 10.1080/14786419.2020.1817014

[68] Wu L, Zhang C, Long Y, Chen Q, Zhang W, Liu G. Food additives: From functions to analytical methods. Critical Reviews in Food Science and Nutrition. 2022;**62**:8497–8517. DOI: 10.1080/10408398.2021.1929823

[69] Embuscado ME. Spices and herbs: Natural sources of antioxidants - A mini review. Journal of Functional Foods. 2015;**18**:811–819. DOI: 10.1016/j.jff.2015.03.005

[70] Nadeem HR, Akhtar S, Ismail T, Qamar M, Sestili P, Saeed W, Azeem M,

Esatbeyoglu T. Antioxidant Effect of Ocimum basilicum Essential Oil and Its Effect on Cooking Qualities of Supplemented Chicken Nuggets. Antioxidants. 2022;**11**:1889. DOI: 10.3390/antiox11101882

[71] Gorinstein S, Martín-Belloso O, Park Y-S, Haruenkit R, Lojek A, Ĉíž M, Caspi A, Libman I, Trakhtenberg S. Comparison of some biochemical characteristics of different citrus fruits. Food Chemistry. 2001;**74**:309–315. DOI: 10.1016/S0308-8146(01)00157-1

[72] Mikołajczak N, Sobiechowska DA, Tańska M. Edible flowers as a new source of natural antioxidants for oxidative protection of cold-pressed oils rich in omega-3 fatty acids. Food Research International. 2020;**134**:109216. DOI: 10.1016/j.foodres.2020.109216

[73] Halvorsen BL, Holte K, Myhrstad MC, Barikmo I, Hvattum E, Fagertun Remberg S, Wold A-B, Haffner K, Baugerød H, Frost Andersen L, et al. A Systematic Screening of Total Antioxidants in Dietary Plants. The Journal of Nutrition. 2002;**132**:461–471

[74] Barros L, Morales P, Carvalho AM, Ferreira ICFR. Antioxidant potential of wild plant foods. Mediterranean Wild Edible Plants. 2016;209–232. DOI: 10.1007/978-1-4939-3329-7_10

[75] Embuscado ME. Herbs and spices as antioxidants for food preservation. Handbook of Antioxidants for Food Preservation. 2015;251–283. DOI: 10.1016/B978-1-78242-089-7.00011-7

[76] Benchagra L, Berrougui H, Islam MO, Ramchoun M, Boulbaroud S, Hajjaji A, Fulop T, Ferretti G, Khalil A. Antioxidant effect of Moroccan pomegranate (Punica granatum L. sefri variety) extracts rich in punicalagin against the oxidative stress process. Foods. 2021;**10**:2219. DOI: 10.3390/foods10092219

[77] Orlo E, Russo C, Nugnes R, Lavorgna M, Isidori M. Natural methoxyphenol compounds: Antimicrobial activity against foodborne pathogens and food spoilage bacteria, and role in antioxidant processes. Foods. 2021;**10**:1807. DOI: 10.3390/foods10081807

[78] Hanafy SM, Abd El-Shafea YM, Saleh WD, Fathy HM. Chemical profiling, in vitro antimicrobial and antioxidant activities of pomegranate, Orange and banana peel-extracts against pathogenic microorganisms. Journal of Genetic Engineering and Biotechnology. 2021;**19**:80. DOI: 10.1186/s43141-021-00151-0

[79] Al-Hashimi AG, Ammar AB, Lakshmanan G, Cacciola F, Lakhssassi N. Development of a millet starch edible film containing clove essential oil. Foods. 2020;**9**:184. DOI: 10.3390/foods9020184

[80] Borotová P, Galovičová L, Vukovic NL, Vukic M, Kunová S, Hanus P, Kowalczewski PŁ, Bakay L, Kačániová M. Role of Litsea cubeba Essential Oil in Agricultural Products Safety: Antioxidant and Antimicrobial Applications. Plants. 2022;**11**:1504. DOI: 10.3390/plants11111504

[81] Wu Y-T, Huo Y-F, Xu L, Xu -Y-Y, Wang X-L, Zhou T. Purification, characterization and antioxidant activity of polysaccharides from Porphyra haitanensis. International Journal of Biological Macromolecules. 2020;**165**:2116–2125. DOI: 10.1016/j.ijbiomac.2020.10.053

[82] Radünz M, Dos Santos Hackbart HC, Camargo TM, Nunes CFP, de Barros FAP, Dal Magro J, Filho PJS, Gandra EA, Radünz A-L, da Rosa Zavareze E. Antimicrobial potential of spray drying encapsulated thyme (Thymus vulgaris) essential oil on the conservation of hamburger-like meat products. International Journal of Food Microbiology. 2020;**330**:108696. DOI: 10.1016/j.ijfoodmicro.2020.108696

[83] Badawy MEI, Lotfy TMR, Shawir SMS. Facile synthesis and characterizations of antibacterial and antioxidant of chitosan monoterpene nanoparticles and their applications in preserving minced meat. International Journal of Biological Macromolecules. 2020;**156**:127–136. DOI: 10.1016/j.ijbiomac.2020.04.044

[84] Wang Z, He Z, Zhang D, Li H. Antioxidant activity of purslane extract and its inhibitory effect on the lipid and protein oxidation of rabbit meat patties during chilled storage. Journal of the Science of Food and Agriculture. 2021;**101**:1953–1962. DOI: 10.1002/jsfa.10811

[85] Abdeldaiem MH. Use of yellow pigment extracted from turmeric (Curcuma longa) rhizomes powder as natural food preservative and colorant. Food Science and Quality Management. 2013;**22**:56–68

[86] Shadordizadeh T, Mahdian E, Hesarinejad MA. Application of encapsulated Indigofera tinctoria extract as a natural antioxidant and colorant in ice cream. Food Science & Nutrition. 2023;**11**:1940–1951. DOI: 10.1002/fsn3.3228

[87] Wahyuningsih S, Ramelan AH, Wardani DK, Aini FN, Sari PL, Tamtama BPN, Kristiawan YR. Indigo Dye Derived from Indigofera tinctoria as Natural Food Colorant. IOP Conference Series: Materials Science and Engineering. 2017;**193**:012048. DOI: 10.1088/1757-899X/193/1/012048

[88] Khaki M, Sahari MA, Barzegar M. Evaluation of Antioxidant and Antimicrobial Effects of Chamomile (Matricaria chamomilla L.) Essential Oil on Cake Shelf Life. Journal of Medicinal Plants. 2012;**11**:9–18

[89] Mahmmod ZA. The effect of chamomile plant (Matericaria chamomile L.) as feed additives on productive performance, carcass characteristics and immunity response of broiler. International Journal of Poultry Science. 2013;**12**:1116. DOI: 10.3923/ijps.2013.111.116

[90] EFSA Panel on Food Additives and Nutrient Sources Added to Food (ANS). Scientific Opinion on the safety of steviol glycosides for the proposed uses as a food additive. EFSA Journal. 2010;**8**:1537. DOI: 10.2903/j.efsa.2010.1537

[91] Milić A, Daničić T, Horecki AT, Šumić Z, Teslić N, Kovačević DB, Putnik P, Pavlić B. Sustainable Extractions for Maximizing Content of Antioxidant Phytochemicals from Black and Red Currants. Foods. 2022;**11**:325. DOI: 10.3390/foods11030325

[92] Meng Q, Chen Z, Chen F, Zhang Z, Gao W. Optimization of ultrasonic-assisted extraction of polysaccharides from Hemerocallis citrina and the antioxidant activity study. Journal of Food Science. 2021;**86**:3082–3096. DOI: 10.1111/1750-3841.15806

[93] Sun C, Wang G, Sun J, Yin J, Huang J, Li Z, Mu D, He M, Liu T,

Cheng J, et al. A New Method of Extracting Polygonatum sibiricum Polysaccharide with Antioxidant Function: Ultrasound-Assisted Extraction-Deep Eutectic Solvents Method. Foods. 2023;**12**:3438

[94] Pavlić B, Mrkonjić Ž, Teslić N, Kljakić AC, Pojić M, Mandić A, Stupar A, Santos F, Duarte ARC, Mišan A. natural deep eutectic solvent (nades) extraction improves polyphenol yield and antioxidant activity of wild thyme (thymus serpyllum L.) extracts. Molecules. 2022;**27**:1508. DOI: 10.3390/molecules27051508

[95] Alzaidi MA, Kumar P, Ismail-Fitry MR, Jusoh S, Ab Aziz MF, Sazili AQ. Green extraction of bioactive compounds from plant biomass and their application in meat as natural antioxidant. Antioxidants. 2021;**10**:1465. DOI: 10.3390/antiox10091465

[96] Milenković A, Aleksovski S, Miteva K, Milenković L, Stanojević J, Nikolić G, Ilić ZS, Stanojević L. The effect of extraction technique on the yield, extraction kinetics and antioxidant activity of black pepper (Piper nigrum L.) ethanolic extracts. Horticulturae. 2023;**11**:125. DOI: 10.3390/horticulturae11020125

[97] Dhanani T, Shah S, Gajbhiye NA, Kumar S. Effect of extraction methods on yield, phytochemical constituents and antioxidant activity of Withania somnifera. Arabian Journal of Chemistry. 2017;**10**:S1193–S1199. DOI: 10.1016/j.arabjc.2014.04.007

[98] Rukavina I, Rodrigues MJ, Pereira CG, Mansinhos I, Romano A, Ślusarczyk S, Matkowski A, Custódio L. Greener is better: First approach for the use of natural deep eutectic solvents (nades) to extract antioxidants from the medicinal halophyte Polygonum maritimum L. Molecules. 2021;**26**:6136. DOI: 10.3390/molecules26206136

[99] Santos F, Soares C, Morais SL, Neves C, Grosso C, Ramalhosa MJ, Vieira M, Delerue-Matos C, Domingues VF. Optimized extraction protocols for bioactive antioxidants from commercial seaweeds in Portugal: A comparative study of techniques. Foods. 2025;**14**:453. DOI: 10.3390/foods14030453

[100] Bastos KVLS, de Souza A, Tomé AC, de Moura Souza F. New strategies for the extraction of antioxidants from fruits and their by-products: A systematic review. Plants. 2025;**14**:755. DOI: 10.3390/plants14050755

[101] Bouizgma K, Rabbah N, Abbas Z, Abourriche A. Unlocking sustainable extraction of natural antioxidants: Green solvents, smart technologies, scalability and future directions. Separation Science and Technology. 2025;**60**:657–683. DOI: 10.1080/01496395.2025.2452411

[102] Peanparkdee M, Patrawart J, Iwamoto S. Physicochemical stability and in vitro bioaccessibility of phenolic compounds and anthocyanins from Thai rice bran extracts. Food Chemistry. 2020;**329**:127157. DOI: 10.1016/j.foodchem.2020.127157

[103] Boulebd H, Carmena-Bargueño M, Pérez-Sánchez H. Exploring the antioxidant properties of caffeoylquinic and feruloylquinic acids: A computational study on hydroperoxyl radical scavenging and xanthine oxidase inhibition. Antioxidants. 2023;**13**:12. DOI: 10.3390/antiox12091669

[104] Rizzardi N, Pezzolesi L, Samorì C, Senese F, Zalambani C, Pitacco W, Calonghi N, Bergamini C, Prata C, Fato R. Natural astaxanthin is a green antioxidant able to counteract lipid peroxidation and ferroptotic cell death. International Journal of Molecular Sciences. 2022;**23**:15137. DOI: 10.3390/ijms232315137

[105] Ghafoor K, Sarker MZI, Al-Juhaimi FY, Babiker EE, Alkaltham MS, Almubarak AK. Extraction and Evaluation of Bioactive Compounds from Date (Phoenix dactylifera) Seed Using Supercritical and Subcritical CO2 Techniques. Foods. 2022;**11**:1806. DOI: 10.3390/foods11121806

[106] Barbosa JR, Freitas MM S, Oliveira LC, S. Martins LH, Almada-Vilhena AO, Oliveira RM, Pieczarka JC, Do S BBD, Carvalho Junior RN. Obtaining extracts rich in antioxidant polysaccharides from the edible mushroom Pleurotus ostreatus using binary system with hot water and supercritical CO2. Food Chemistry. 2020;**330**:127173. DOI: 10.1016/j.foodchem.2020.127173

[107] Li J, Niu D, Zhang Y, Zeng X-A. Physicochemical properties, antioxidant and antiproliferative activities of polysaccharides from Morinda citrifolia L. (Noni) based on different extraction methods. International Journal of Biological Macromolecules. 2020;**150**:114–121. DOI: 10.1016/j.ijbiomac.2019.12.157

[108] Wan C, Jiang H, Tang M-T, Zhou S, Zhou T. Purification, physico-chemical properties and antioxidant activity of polysaccharides from Sargassum fusiforme by hydrogen peroxide/ascorbic acid-assisted extraction. International Journal of Biological Macromolecules. 2022;**223**:490–499. DOI: 10.1016/j.ijbiomac.2022.11.030

[109] Rodríguez De Luna SL, Ramírez-Garza RE, Serna Saldívar SO. Environmentally friendly methods for flavonoid extraction from plant material: Impact of their operating conditions on yield and antioxidant properties. The Scientific World Journal. 2020;**2020**:6792069. DOI: 10.1155/2020/6792069

[110] Rwubuzizi R, Kim H, Holzapfel WH, Todorov SD. Beneficial, safety, and antioxidant properties of lactic acid bacteria: A next step in their evaluation as potential probiotics. Heliyon. 2023;**9**:e15610. DOI: 10.1016/j.heliyon.2023.e15610

[111] Sun SJ, Deng P, Peng CE, Ji HY, Mao LF, Peng LZ. Extraction, structure and immunoregulatory activity of low molecular weight polysaccharide from Dendrobium officinale. Polymers. 2022;**14**(2899):2899. DOI: 10.3390/polym14142899

[112] Scassellati C, Galoforo AC, Bonvicini C, Esposito C, Ricevuti G. Ozone: A natural bioactive molecule with antioxidant property as potential new strategy in aging and in neurodegenerative disorders. Ageing Research Reviews. 2020;**63**:101138. DOI: 10.1016/j.arr.2020.101138

[113] Chernukha I, Fedulova L, Vasilevskaya E, Kulikovskii A, Kupaeva N, Kotenkova E. Antioxidant effect of ethanolic onion (Allium cepa) husk extract in ageing rats. Saudi Journal of Biological Sciences. 2021;**28**:2877–2885. DOI: 10.1016/j.sjbs.2021.02.020

[114] El Sayed NS, Ghoneum MH. Antia, a Natural Antioxidant Product,

Attenuates Cognitive Dysfunction in Streptozotocin-Induced Mouse Model of Sporadic Alzheimer's Disease by Targeting the Amyloidogenic, Inflammatory, Autophagy, and Oxidative Stress Pathways. Oxidative Medicine and Cellular Longevity. 2020;**2020**:1–14. DOI: 10.1155/2020/4386562

[115] Hua Z, Zhang X, Zhao X, Zhu B-W, Liu D, Tan M. Hepatic-targeted delivery of astaxanthin for enhanced scavenging free radical scavenge and preventing mitochondrial depolarization. Food Chemistry. 2023;**406**:135036. DOI: 10.1016/j.foodchem.2022.135036

[116] Ali AA, Abd El-Fattah AI, Abu-Elfotuh K, Elariny HA. Natural antioxidants enhance the power of physical and mental activities versus risk factors inducing progression of Alzheimer's disease in rats. International Immunopharmacology. 2021;**96**:107729. DOI: 10.1016/j.intimp.2021.107729

[117] Daniel PS, Lourenço ELB, Sete da Cruz RM, De Souza Gonçalves CH, Das Almas LRM, Hoscheid J, da Silva C, Jacomassi E, Brum Junior L, Alberton O. Composition and antimicrobial activity of essential oil of yarrow (Achillea millefolium L.). Australian Journal of Crop Science. 2020;**14**:545–550. DOI: 10.21475/ajcs.20.14.03.p2325

[118] Rodrigues CRF, Dias JH, de Mello RN, Richter MF, Picada JN, Ferraz ABF. Genotoxic and antigenotoxic properties of Baccharis trimera in mice. Journal of Ethnopharmacology. 2009;**125**:97–101. DOI: 10.1016/j.jep.2009.06.006

[119] Ghaemi H, Masoompour SM, Afsharypuor S, Mosaffa-Jahromi M, Pasalar M, Ahmadi F, Niknahad H. The effectiveness of a traditional Persian medicine preparation in the treatment of chronic cough: A randomized, double-blinded, placebo-controlled clinical trial. Complementary Therapies in Medicine. 2020;**49**:102324. DOI: 10.1016/j.ctim.2020.102324

[120] Mendes TC, da Silva GR, Silva AO, Schaedler MI, Guarnier LP, Palozi RAC, Signor CT, Dal Bosco JD, Auth PA, Amaral EC, Froelich DL, Soares AA, Lovato ECW, Ribeiro-Paes JT, Gasparotto Junior A, Dos Reis Lívero FA. Hepato- and cardioprotective effects of Baccharis trimera (Less.) DC. against multiple risk factors for chronic noncommunicable diseases. Anais da Academia Brasileira de Ciências. 2021;**93**:e202020899. DOI: 10.1590/0001-3765202120200899

[121] Cechinel-Zanchett CC, Fonseca da Silva RCMA, Tenfen A, Siebert DA, Micke G, Vitali L, Cechinel-Filho V, Faloni de Andrade S, de Souza P. Bauhinia forficata link, a Brazilian medicinal plant traditionally used to treat cardiovascular disorders, exerts endothelium-dependent and independent vasorelaxation in thoracic aorta of normotensive and hypertensive rats. Journal of Ethnopharmacology. 2019;**243**:112118. DOI: 10.1016/j.jep.2019.112118

[122] Maideen NMP. Probable beneficial effects of black seeds (Nigella sativa) in the management of langya henipavirus infection. Food and Health. 2022;**4**:17. DOI: 10.53388/FH20221101017

[123] Shahidi F, Zhong HYJ, Ambigaipalan P. Antioxidants: Regulatory Status. John Wiley & Sons, Ltd; 2020. p. 491–512. DOI: 10.1002/047167849X.bio035.pub2

[124] Salami SA, Guinguina A, Agboola JO, Omede AA, Agbonlahor EM, Tayyab U. Review: In vivo and postmortem effects of feed antioxidants in livestock: A review of the implications on authorization of antioxidant feed additives. Animal. 2016;**10**:1375-1390. DOI: 10.1017/S1751731115002967

[125] Mohammadi N, Dos Santos Lima A, Azevedo L, Granato D. Bridging the gap in antioxidant activity of flavonoids: Correlating the oxidation of human plasma with chemical and cellular assays. Current Research in Food Science. 2024;**8**:100714. DOI: 10.1016/j.crfs.2024.100714

[126] Platzer M, Kiese S, Herfellner T, Schweiggert-Weisz U, Miesbauer O, Eisner P. Common trends and differences in antioxidant activity analysis of phenolic substances using single electron transfer based assays. Molecules. 2021;**26**:1244. DOI: 10.3390/molecules26051244

[127] Chaves N, Santiago A, Alías JC. Quantification of the antioxidant activity of plant extracts: Analysis of sensitivity and hierarchization based on the method used. Antioxidants. 2020;**9**:76. DOI: 10.3390/antiox9010076

[128] Liu Y, Xiao Z, Chen F, Yue L, Zou H, Lyu J, Wang Z. Metallic oxide nanomaterials act as antioxidant nanozymes in higher plants: Trends, meta-analysis, and prospect. Science of the Total Environment. 2021;**780**:146578. DOI: 10.1016/j.scitotenv.2021.146578

Chapter 4

Tartaric Instability in Wine: Mechanisms and Additive-Based Stabilization Strategies

Fernanda Cosme, Juliana Milheiro, Luís Moreira, Sandrine Ferreira, Luís Filipe-Ribeiro, Mário Bezerra, Miguel Ribeiro and Fernando M. Nunes

Abstract

Tartaric acid is usually the main organic acid in wine; however, it readily forms poorly soluble salts that can precipitate as crystals. The presence of crystalline precipitates in bottled wine is frequently perceived by consumers as a quality defect. These precipitates result from tartaric instability, a natural physicochemical phenomenon mainly driven by the crystallization of potassium bitartrate and, to a lesser extent, calcium tartrate in supersaturated wine systems. The occurrence of tartrate crystallization is influenced by several physicochemical factors, including pH, temperature, alcohol content, ionic composition, and natural colloids. Consequently, tartaric stability is a key quality parameter in wine production, and controlling it is essential in winemaking. In response to economic, technological, and environmental pressures, the wine industry has increasingly adopted additive-based strategies for tartaric stabilization. These treatments inhibit crystal nucleation and growth without removing wine constituents. This chapter provides an overview of the mechanisms underlying tartaric instability in wines, including its chemical basis and physicochemical factors influencing crystal nucleation and growth. It also reviews methods used to assess tartaric stability prior to bottling, including traditional cold tests, the conductimetric minicontact test, and spectroscopic techniques based on Fourier transform infrared (FTIR) spectroscopy combined with multivariate analysis. Particular attention is given to additive-based tartaric stabilization strategies authorized for enological use, such as metatartaric acid, carboxymethylcellulose, mannoproteins, potassium polyaspartate, and gum arabic. Their mechanisms of action, technological effectiveness, interactions with wine components, and sensory implications are discussed Regulatory frameworks governing their use and legal limits established by international authorities are also addressed.

Keywords: tartaric instability, potassium bitartrate, calcium tartrate, enological additives, carboxymethylcellulose, metatartaric acid, potassium polyaspartate, mannoproteins, gum arabic, conductimetric tests

1. Introduction

The visual limpidity of a wine is the consumer's first point of contact with its perceived quality. While slight cloudiness may be acceptable in certain artisanal styles, the presence of crystalline precipitates is frequently misinterpreted by consumers as glass fragments or chemical contamination [1]. These crystals arise from tartaric instability, a natural physicochemical phenomenon primarily involving the precipitation of potassium bitartrate (KHT) and, to a lesser extent, calcium tartrate (CaT) [2, 3].

Tartaric acid is the key organic acid of the grapevine (*Vitis vinifera*), providing essential structural acidity and pH control that are critical for microbial stability and sensory balance in wine [4]. However, wine is a complex hydroethanolic solution that is often supersaturated with respect to tartrate salts [2]. As wine ages or is exposed to low temperatures during transport and storage, this supersaturation can reach a critical threshold, triggering the nucleation and growth of tartrate crystals. For decades, the industry-standard approach to preventing this phenomenon has been "cold stabilization," a subtractive process in which wine is chilled to near-freezing temperatures to induce tartrate precipitation before bottling. Although effective, cold stabilization is energy-intensive and time-consuming and may, in some cases, lead to losses of aromatic compounds or a reduction in color intensity [3].

In contrast to KHT, the initial nucleation of CaT is not effectively triggered by temperature reduction or by increasing the degree of supersaturation in the wine matrix. Consequently, cold stabilization is generally ineffective for CaT stabilization [5].

Enology practices and research have increasingly explored additive-based stabilization as an alternative strategy. Unlike subtractive methods, these additives act as "crystallization inhibitors" or "protective colloids," interfering with crystal nucleation and growth. This approach allows wine to maintain visual stability under adverse conditions without the need for extreme refrigeration [6].

This chapter provides a comprehensive overview of additives used for tartaric stabilization in wine (see **Figure 1**). It begins with an examination of the chemical and technological mechanisms underlying tartaric instability, followed by a review of the analytical techniques employed to evaluate stabilization requirements. The core of the chapter focuses on the principal additives authorized for tartaric stabilization [7], including metatartaric acid (MTA) [8, 9], carboxymethylcellulose (CMC) [10, 11], mannoproteins (MP), potassium polyaspartate (KPA) [12–14], and gum arabic (GA) [6, 14–16]. Particular attention is given to their mechanisms of action, technological effectiveness, and sensory impact. The regulatory frameworks governing their use and the legal limits established by international authorities are also discussed. Finally, the chapter considers emerging trends in sustainability and innovation, highlighting the role of additive-based stabilization in supporting environmentally responsible winemaking practices and shaping future developments in this field.

2. Tartaric instability in wine

Tartaric acid is the major organic acid in grapes, grape must, and wine, with typical concentrations ranging from 2 to 10 g/L [4]. In wine, tartaric acid exists in

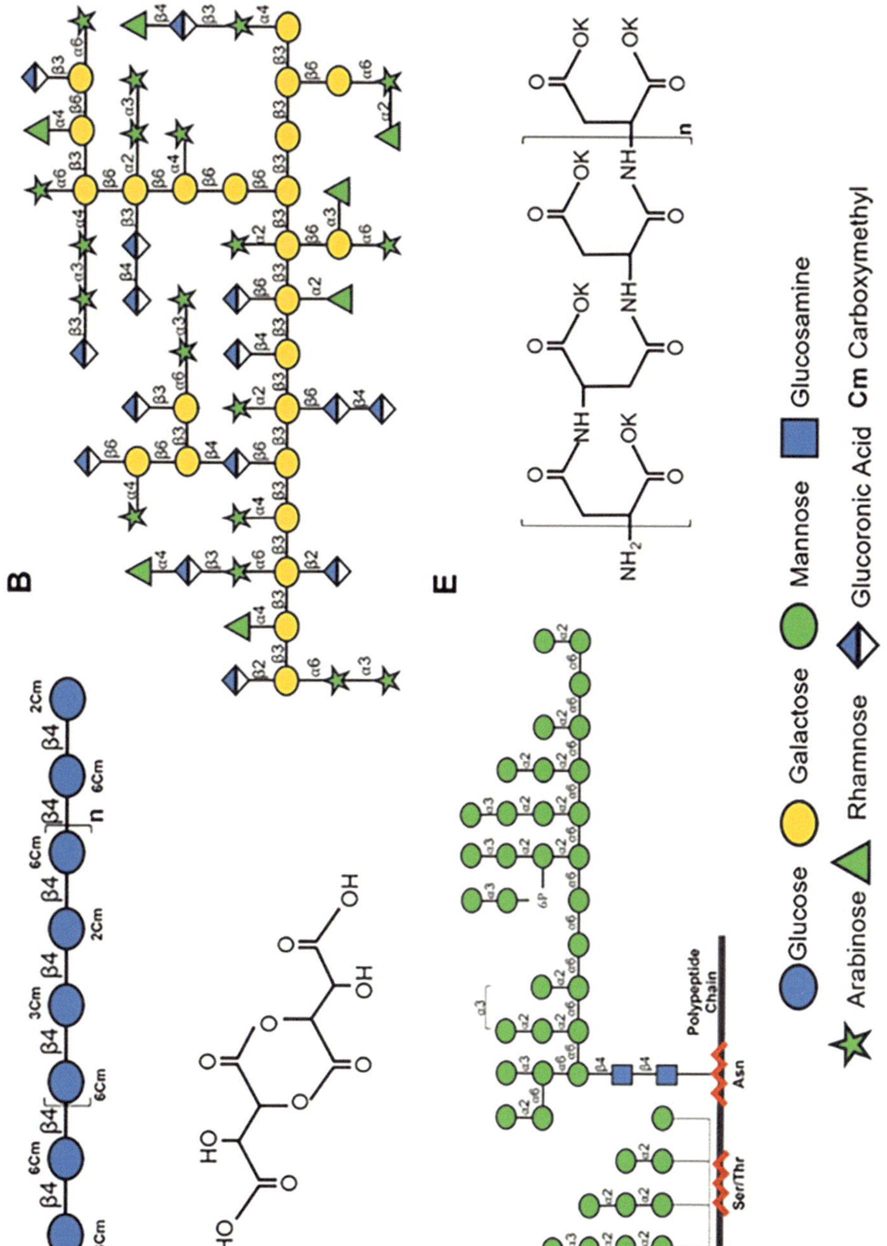

Figure 1.
Additives authorized by the OIV for tartaric stabilization in wine: (a) CMC, (b) GA from Acacia senegal*, (c) MTA, (d)* Saccharomyces cerevisiae *mannoprotein, and (e) KPA. Polysaccharide structures were drawn using the symbol nomenclature for graphical representation of glycans. Source: From literature [7, 17, 18, 19].*

three chemical forms: free tartaric acid (H_2T), hydrogen tartrate (HT^-), and tartrate (T^{2-}), whose relative distribution depends primarily on pH but is also influenced by temperature, ionic strength, and ethanol concentration [20]. Higher alcohol content, lower temperature, and higher pH result in decreased KHT solubility [1]. Within the usual wine pH range, HT^- predominates, accounting for approximately 50–70% of the total tartaric acid.

$$H_2T \leftrightarrow HT^- + H^+ \leftrightarrow T^{2-} + H^+ \tag{1}$$

2.1 Crystallization mechanisms and tartaric instability in wine

Tartaric instability in wine arises from the formation of insoluble tartrate salts, primarily KHT and CaT. The formation of tartrate crystals is a complex physicochemical process involving crystal nucleation followed by crystal growth. Nucleation may occur through homogeneous mechanisms, whereby crystal-forming ions spontaneously associate in a supersaturated solution through electrostatic interactions, or through heterogeneous mechanisms, in which suspended solid particles act as nucleation sites [21]. Once stable nuclei are formed, crystal growth proceeds through the diffusion and incorporation of ions onto the crystal surface, ultimately leading to the precipitation of tartrate salts.

From a thermodynamic perspective, tartrate precipitation is governed by the solubility equilibria of KHT and CaT. When the concentrations of tartaric acid and cations, such as potassium (1.5–2.5 g/L) and/or calcium (40–170 mg/L), exceed their respective solubility products, the wine becomes supersaturated, creating favorable conditions for crystallization [3, 22]. The dissociation and precipitation equilibria of these salts can be described by the following reactions:

$$KHT(s) \leftrightharpoons HT^-(aq) + K^+(aq) \tag{2}$$

$$CaT(s) \leftrightharpoons T^{2-}(aq) + Ca^{2+}(aq) \tag{3}$$

The solubility product constant (Ksp) defines the equilibrium between the solid phase and dissolved ions. For KHT, $Ksp = [HT^-][K^+] = 3.8 \times 10^{-4}$ at 18 °C, whereas for CaT, $Ksp = [T^{2-}][Ca^{2+}] = 7.7 \times 10^{-7}$ at 25 °C. When the ionic product equals Ksp, the system is at saturation, and no crystallization occurs. Under undersaturated conditions, additional solute can dissolve; however, when the ionic product exceeds Ksp, the system becomes supersaturated, triggering nucleation and subsequent crystal growth, which ultimately results in tartrate instability [23].

Crystal formation proceeds through three sequential stages: supersaturation, nucleation, and growth. Supersaturation is a necessary but not sufficient condition for nucleation, as the formation of stable nuclei requires overcoming a significant energetic barrier. Primary nucleation occurs spontaneously in highly supersaturated wines or may be induced by external factors such as temperature reduction, generally producing fewer but larger crystals and exhibiting longer induction times. Secondary, or induced, nucleation may occur homogeneously through the presence of crystals of the same chemical nature already present in the wine or intentionally added, or heterogeneously via solid particles of different composition acting as nucleation sites [4].

Once nucleation has occurred, crystal growth is controlled by the diffusion and incorporation of K^+ and HT^- ions or Ca^{2+} and T^{2-} ions onto active sites on the crystal surface. Owing to the higher affinity of K^+ ions relative to HT^- ions, crystal surfaces tend to acquire a net positive charge, favoring the adsorption of negatively charged colloidal particles. The accumulation of such colloids on the crystal surface can hinder further growth and modulate precipitation kinetics [24]. Crystal growth continues until thermodynamic equilibrium is restored and the ionic product falls below the solubility threshold.

Because potassium concentrations in wine are typically an order of magnitude higher than those of calcium, KHT is the dominant crystalline precipitate observed in wines, making it the principal contributor to tartaric instability under most winemaking conditions [25].

2.2 Factors affecting tartaric stability in wine

2.2.1 Ionic composition and tartrate species

The concentrations of potassium (K^+), calcium (Ca^{2+}), and the conjugate bases of tartaric acid (HT^- and T^{2-}) play a central role in tartrate equilibrium. Potassium and calcium ions interact with hydrogen tartrate and tartrate ions to form KHT and CaT, respectively. When wine reaches a supersaturated state, excess tartrate precipitates and crystallizes. Conversely, a decrease in the concentration of any of these ions shifts the equilibrium toward dissolution, reducing the likelihood of precipitation [26]. Therefore, controlling the ionic composition is a key strategy for managing tartaric stability.

2.2.2 Colloidal substances

Wine contains numerous colloidal components, often referred to as protective colloids, including proteins, phenolic compounds, and polysaccharides, which significantly influence tartrate crystallization behavior [1]. These compounds exert an inhibitory or retarding effect on the crystallization of KHT and CaT by limiting ion migration toward crystal nuclei and by adsorbing onto crystal surfaces, thereby blocking the incorporation of additional ions and slowing crystal growth [27, 28].

Proteins are particularly effective at inhibiting crystal growth due to their ability to adsorb onto tartrate crystal faces and form a protective layer. At typical wine pH values, partial protein unfolding increases the number of exposed binding sites, enhancing this inhibitory effect. However, protein–tannin interactions introduce additional complexity. At low protein-to-tannin ratios, protein–tannin aggregates may precipitate, diminishing the protective action of proteins and facilitating tartrate crystallization. At higher tannin concentrations, tannins can interact directly with crystal surfaces through hydrogen bonding; nevertheless, electrostatic repulsion at low pH may weaken these interactions, reducing their stabilizing effectiveness [29].

Polysaccharides such as pectic substances, rhamnogalacturonans (RG-I and RG-II), arabinogalactan proteins, and MPs also contribute significantly to tartrate stability. Uronic acid–rich polysaccharides, particularly RG-I, are among the most effective inhibitors of CaT precipitation due to their ability to chelate Ca^{2+} ions and delay nucleation. The RG-II exhibits a similar but less pronounced effect, while MPs generally show limited inhibitory activity at wine pH due to their reduced calcium-binding capacity. The lower concentration of such polysaccharides in white wines partly explains their higher susceptibility to tartrate precipitation compared with red wines [30, 31].

2.2.3 Organic acids

Several organic acids naturally present in wine play an important role in modulating tartaric stability by competing with tartaric acid for potassium (K^+) and calcium (Ca^{2+}) ions, thereby inhibiting tartrate nucleation. Among these, citric and malic acids are particularly effective inhibitors of CaT crystallization, whereas lactic and succinic acids exhibit weaker effects. This distinction is particularly relevant during alcoholic and malolactic fermentation, as microbial metabolism progressively converts malic and citric acids into lactic and succinic acids, resulting in a net reduction in cation-complexing capacity.

During malolactic fermentation, the conversion of malic acid to lactic acid is accompanied by an increase in wine pH and a decrease in the concentration of effective crystallization inhibitors. These combined effects promote enhanced supersaturation with respect to tartrate salts and increase the risk of tartrate precipitation in finished wines. This phenomenon is especially critical in sparkling wines and full-bodied white wines, where the lower abundance of protective colloids further exacerbates tartaric instability. In contrast, red wines generally achieve tartaric stability more readily, owing to their higher content of phenolic compounds and other protective colloids that mitigate crystal nucleation and growth [3, 32].

2.3 pH

pH strongly influences the dissociation state of tartaric acid and, consequently, tartrate stability. As pH increases, the relative proportions of HT^- and T^{2-} change, affecting the availability of species involved in salt formation. Wines with higher pH values generally exhibit increased concentrations of ionic tartrate species, which enhances the risk of tartrate precipitation [20].

2.3.1 Temperature

Temperature exerts a strong influence on the solubility and crystallization behavior of tartrate salts in wine. Increasing temperature enhances the solubility of KHT and CaT, favoring crystal dissolution, whereas decreasing temperature reduces solubility and promotes crystallization. However, the solubility of CaT exhibits markedly lower temperature dependence than that of KHT. Although low temperatures can accelerate crystallization kinetics, cooling alone, even to temperatures approaching the freezing point of wine, does not necessarily induce CaT precipitation.

In many cases, despite increased levels of CaT supersaturation, crystal nucleation does not occur, indicating that supersaturation is a necessary but insufficient condition for CaT crystallization [20, 33]. In contrast, KHT crystallization is more readily promoted by cooling, particularly in the presence of suitably sized particles, as crystal growth and precipitation are strongly influenced by surface area effects. Particles smaller than 50 μm have been shown to be especially effective in facilitating KHT precipitation [34]. This differential temperature sensitivity underpins the widespread use of cold storage and cold stabilization treatments to selectively induce tartrate precipitation prior to bottling.

2.3.2 Alcohol content

Alcohol content plays a critical role in the solubility and stability of tartrate salts in wine. This behavior is primarily attributed to the lower polarity of ethanol compared to water, which reduces the solvent capacity of the hydroalcoholic medium for polar and ionic compounds, including tartrate salts. As a result, KHT exhibits significantly lower solubility in alcoholic solutions. For example, in wines containing 12–13% (v/v) ethanol, approximately half of the tartrate salts may precipitate relative to their initial dissolved content, reflecting the reduced solubilizing capacity associated with higher alcohol levels.

Consistently, experimental studies have shown that increasing ethanol concentrations raises the saturation temperature of tartrate salts and intensifies precipitation tendencies, thereby reducing overall tartrate stability in wine [23, 29].

3. Analytical methods for assessing tartaric stability

Evaluating the tartaric stability of wine is essential for managing stabilization practices and assessing the effects of additives. Some analytical methods have been developed to estimate the risk of tartrate precipitation prior to bottling, ranging from empirical crystallization tests to predictive approaches based on spectroscopic analysis and conductimetry.

The cold test has traditionally been used as a reference method for assessing tartaric stability. In this approach, wine samples are stored at a low temperature, typically around -4 °C, for several days to promote KHT crystallization. Instability is then evaluated either by visual observation of precipitates [35] or by quantitative methods, such as gravimetric determination or measurement of tartaric acid loss after cold filtration, often using high-performance liquid chromatography (HPLC) [26]. The cold test is time-consuming and frequently considered overly severe, as wines deemed unstable by this method may remain clear under long-term storage at higher, commercially relevant temperatures [36]. Moreover, it reflects both primary nucleation and secondary crystal growth phenomena, limiting its ability to isolate specific mechanisms of instability.

To obtain faster and more operational assessments, minicontact or conductivity tests have been developed to reproduce tartaric precipitation through induced secondary crystallization. These methods involve the addition of KHT crystals to wine under controlled temperature and agitation. The extent of crystallization can be monitored by weight loss [37], potassium depletion, or changes in electrical conductivity [22], the latter being particularly practical for routine analysis. Among these, the *degré d'instabilité tartrique* (DIT) test measures the conductivity drop over four hours at -4 °C after the addition of 4 g/L KHT crystals, with stability thresholds initially set at 3% and later increased to 5% [38]. This conductimetric approach was originally designed to estimate the target conductivity value for tartaric stabilization by electrodialysis; however, it has proven unsuitable for evaluating residual tartaric stability after such treatment.

To address this limitation, the *critical stability index (ISTC50)* was introduced [39]. In this method, 0.5 g/L KHT is dissolved in wine under stirring at 36 °C, followed by a controlled decrease to 4 °C, while conductivity is continuously monitored. The degree of tartaric instability is determined from the induction time preceding the abrupt conductivity drop associated with crystal formation.

A methodological advancement within this framework was proposed by Bosso et al. [40], who developed and validated a rapid minicontact conductimetric test for evaluating wine tartaric stability. This method measures the conductivity drop in wine maintained under stirring at 0 °C for a short contact period following the addition of a standardized dose of micronized KHT crystals. By systematically optimizing temperature, crystal dose, contact time, and crystal size, the method significantly improves repeatability and reproducibility. It allows rapid discrimination between stable and unstable wines and establishes distinct stability thresholds for white and red/rosé wines, reflecting compositional differences that influence crystallization kinetics.

In parallel with empirical crystallization tests, thermodynamic and compositional approaches have been proposed to estimate tartaric stability based on parameters such as alcohol content, pH, ionic strength, and potassium and tartaric acid concentrations [20, 34]. Other methods determine the saturation temperature of KHT as an indicator of supersaturation [34]. However, these approaches often fail to accurately predict actual precipitation behavior, as they do not account for the inhibitory effects of wine colloids, including proteins, polysaccharides, and polyphenols, which can induce metastable supersaturated states [41].

More recently, Fourier transform infrared (FTIR) spectroscopy combined with multivariate statistical analysis has been proposed as a rapid, nondestructive method for predicting tartaric instability. Malacarne et al. [42] developed predictive models based on FT-IR spectra of wines using partial least squares regression and artificial neural networks calibrated against reference methods such as the minicontact test and the cold test. This approach enables reliable classification of wines according to their risk of tartrate precipitation for both white and red wines, while considerably reducing analysis time and sample manipulation. The method extends previous applications of infrared spectroscopy in wine analysis to the specific prediction of tartaric instability and represents a promising alternative to conventional empirical testing [42, 43].

4. Authorized additives for wine tartaric stabilization

The use of additives to prevent tartrate crystallization has gained increasing interest as a simple, effective, and economically viable alternative to conventional tartaric stabilization techniques [44]. Compared with other stabilization methods, the use of protective colloids to stabilize tartrate salts in commercial wines is economically advantageous, as it involves low energy consumption and does not require specialized equipment [3]. These approaches, approved by the International Organization of Vine and Wine (OIV), include MTA, carboxymethyl cellulose (CMC), MPs, KPA, and GA (see **Figure 1**) [7, 45]. The application of these additives must be carefully adjusted according to several factors, including wine type, intended shelf life, and the degree of tartaric instability determined through stability testing (see **Figure 2**).

4.1 Metatartaric acid

Metatartaric acid (E353) is produced through the thermal polymerization of tartaric acid under vacuum at high temperatures (150–170 °C), involving

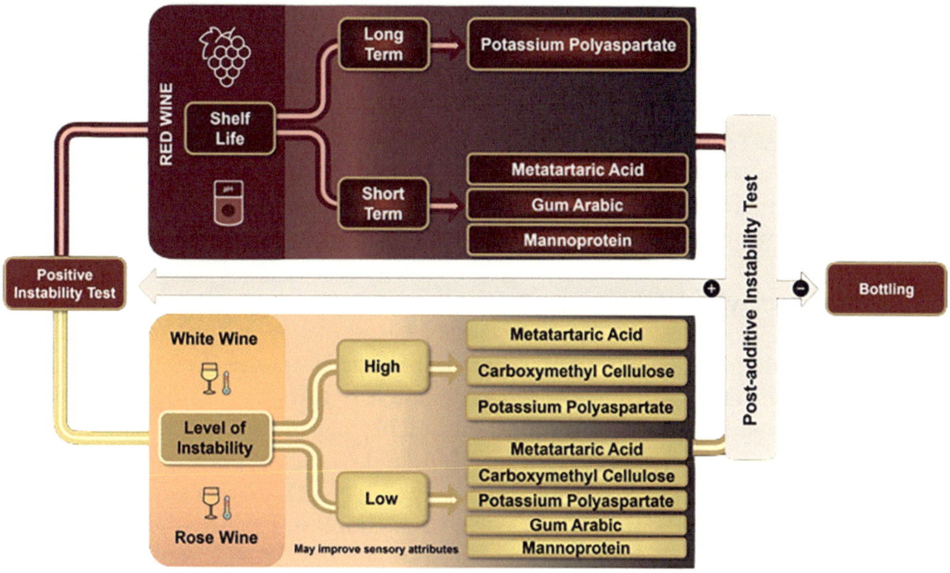

Figure 2.
Decision-making flowchart summarizes the use of additives for tartaric stabilization according to wine type, intended shelf life, and degree of instability.

intramolecular dehydration and esterification reactions. This process yields a polymeric, yellowish-brown material with no well-defined chemical structure and an esterification index of approximately 30–40% [8, 9]. After grinding to appropriate particle sizes, MTA is commercialized in various formulations for wine stabilization. Nevertheless, the production process also generates additional compounds, such as ditartaric monoester and diesters in variable proportions, nonesterified tartaric acid, pyruvic acid, and low amounts of polyester acids [7]. Consequently, commercial MTA products may vary considerably in their degrees of esterification and may contain considerable amounts of monomeric tartaric acid. Furthermore, the maximum permitted dosage of MTA in wine is limited to 100 mg/L, and such compositional variations can significantly influence its stabilizing effectiveness [8].

MTA is highly soluble in water and ethanol, and its hydrolysis is strongly temperature-dependent. Hydrolysis occurs rapidly at high temperatures but is significantly slowed at lower temperatures, thus preserving its stabilizing effect during cool storage [9, 45].

In winemaking practice, MTA is typically applied after final filtration and immediately prior to bottling [45]. At this stage, it acts as an effective inhibitor of KHT and CaT crystal formation, preventing their precipitation while exerting minimal impact on the sensory characteristics of the wine. Additionally, MTA is easy to apply and relatively inexpensive to produce [8].

Despite its broad applicability across different wine types, MTA exhibits limited long-term stability. It provides short- to medium-term protection against tartrate instability, which is generally sufficient for white wines intended for rapid market release. However, its effectiveness declines during prolonged storage because MTA is gradually hydrolyzed to tartaric acid in wine. The rate of hydrolysis depends on both pH and temperature and may range from approximately one week at 30 °C to up to two years at 0 °C, while at typical cellar temperatures (10–18 °C), it generally ranges from about one year to 18 months [3]. Nevertheless, studies have shown that in red wines stored at 20 °C, the stabilizing effect may be lost after approximately five months due to MTA hydrolysis, whereas white wines remained stable for up to 12 months under the same conditions [46]. Therefore, MTA is most suitable for wines with short commercial lifespans and rapid market release, particularly when low storage temperatures are maintained.

4.2 Carboxymethyl cellulose

Carboxymethyl cellulose, a plant-derived polymer, is synthesized through a two-step process: the alkali treatment of wood cellulose to form an alkali–cellulose complex, followed by etherification with monochloroacetic acid [11]. Its low production cost, reduced energy consumption, and ease of application align well with the requirements of low-carbon processes [3]. The CMC was authorized for enological use by the OIV in 2009. In 2023, the maximum permitted dose was updated to 200 mg/L for white, rosé, and sparkling wines [7, 45]. According to OIV Resolution OIV/OENO 366-2009, CMC used for enological applications must be produced exclusively from wood cellulose [7, 45].

Its structural characteristics are defined by the degree of substitution (DS), which corresponds to the proportion of glucose units substituted with carboxymethyl groups, and the degree of polymerization (DP), which represents the average number of glucose units per polymer chain [6]. For enological use, DS must range between 0.60 and 0.95, and DP must range between 80 and 1,500 [7]. With a pKa of approximately 4.3, CMC is partially dissociated at wine pH, and a fraction of its carboxymethyl groups are negatively charged in solution.

CMC enhances tartaric stability by inhibiting both nucleation and crystal growth through a protective colloidal effect. At wine pH, it behaves as a negatively charged polymer that adsorbs onto KHT crystal surfaces, reducing their growth rate and modifying crystal morphology. It can also interact with K^+ (and Ca^{2+}) ions, decreasing the concentration of free cations available for crystallization [47–49]. Higher DS values generally improve its effectiveness by increasing the number of available cation-binding sites [6, 50].

Several studies have demonstrated the positive effect of CMC on tartaric stabilization in wine, showing that its addition provides a stabilizing effect comparable to that of MTA [49, 51, 52]. Furthermore, CMC has been shown to maintain its inhibitory efficiency for up to two months after exposure to 30 °C, unlike MTA, which undergoes hydrolysis under the same conditions [53]. CMC is considered economically viable, easy to apply, and chemically stable, remaining effective during extended storage periods, even at elevated temperatures [47, 49]. No significant changes in pH have been reported compared with treatments using MTA [54]. Sensory studies indicate that, at the maximum dose authorized by the OIV, CMC does not affect taste or aroma, and even substantially higher concentrations have shown no detectable sensory impact [55].

Currently, its use is restricted to white, rosé, and sparkling wines [7, 45], as its effectiveness in red wines may be reduced and, in some cases, has been associated with color loss or turbidity [3, 39, 56, 57]. Color changes in red wines appear to be linked to interactions with proteins involved in turbidity formation [57] or with phenolic compounds[56]. Although CMC can inhibit KHT crystal growth in red wines, certain wine matrices may develop protein haze [57]. Some studies suggest that color loss requires the presence of proteins, possibly due to anthocyanin precipitation mediated by protein bridging [58]. Nevertheless, other reports indicate that CMC can improve tartaric stability in red wines without significantly affecting phenolic composition, sensory attributes, or color stability, depending on wine composition [10, 11].

4.3 Mannoproteins

Mannoproteins are complex glycoconjugates located in the outer layer of yeast cell walls and are composed of mannose-rich polysaccharide chains covalently linked to a protein backbone. During alcoholic fermentation and aging on lees, *Saccharomyces cerevisiae* releases small amounts of these macromolecules into wine as a result of enzymatic autolysis and cell wall remodeling. Although their natural concentration in finished wines remains relatively low, MPs have attracted considerable interest in enology because of their technological properties, particularly their contribution to wine tartaric stabilization [59].

Unlike traditional cold stabilization, which removes tartrate salts from wine, MPs generally act through a crystallization–inhibition mechanism. They may adsorb onto crystal nuclei and block active growth sites, thereby limiting further crystal development [60]. Their amphiphilic structure can also provide colloidal protection, helping to maintain salts in a metastable supersaturated state and reducing the likelihood of aggregation. As a result, wines can remain clear even under cold storage conditions that would normally induce crystallization [61].

The stabilizing efficacy of MPs depends largely on both dosage and the composition of the wine matrix. In white wines, additions of around 250 mg/L have been reported to minimize potassium loss and provide high tartaric stability [57], whereas in red wines, the effect is often less pronounced due to the natural abundance of protective colloids such as polyphenols and polysaccharides [60]. Furthermore, fining agents commonly used in red winemaking, including bentonite and gelatin, can remove MPs from the wine, thereby reducing their stabilizing effectiveness. For this reason, MPs are mainly applied to improve tartaric stability in white and rosé wines [62].

In practical winemaking applications, MPs are generally added after fermentation or before bottling, often following filtration. White and rosé wines are particularly suitable candidates because their relatively low phenolic content makes them more susceptible to tartrate precipitation. Appropriate additions can provide stability comparable to conventional chilling treatments while preserving wine composition [63]. Aromatic varieties such as Sauvignon Blanc or Muscat may particularly benefit from this approach because cold stabilization can lead to losses of volatile compounds through co-precipitation and subsequent filtration, whereas MPs help preserve aromatic intensity while ensuring stability [64, 65].

Compared with conventional stabilization techniques, MP addition offers several advantages. Because tartrate salts remain dissolved rather than removed, the natural acidity and overall balance of the wine tend to be preserved. These benefits are consistent with current trends toward sustainable and low-intervention winemaking. In addition to tartaric stabilization, MPs may also contribute positively to mouthfeel and overall sensory quality, further supporting their technological value in enology [65, 66].

4.4 Potassium polyaspartate

Potassium polyaspartate is synthesized through the thermal polymerization of L-aspartic acid, which leads to the formation of polysuccinimide, an insoluble intermediate with a relatively low molecular weight. This intermediate is subsequently hydrolyzed and neutralized with potassium hydroxide under controlled conditions to produce KPA [7, 45]. Commercial KPA is supplied as a light-brown, odorless powder containing approximately 90% dry matter. It is highly soluble in water (>1,000 g/L) and insoluble in organic solvents (<5 g/L), with a shelf life of up to four years at room temperature [7, 45].

This polymer contains a high number of carboxyl groups that are negatively charged at wine pH. These functional groups enable KPA to interact electrostatically with potassium ions. Through this mechanism, KPA inhibits the nucleation and growth of KHT crystals, thereby preventing tartrate precipitation without removing wine constituents [13, 14].

KPA provides a tartaric stabilization efficiency comparable to that of other additives, while exhibiting greater persistence over time. Bosso et al. [12] demonstrated that KPA is a highly effective and durable stabilizer against KHT precipitation in both white and red wines. Unlike MTA, its stabilizing effect remains effective over extended storage periods without negatively affecting turbidity, color, or phenolic composition. Overall, KPA represents a reliable alternative for tartrate stabilization, particularly suited to wines requiring long-term stability.

In white wines, experimental trials conducted at different storage temperatures (4 °C and 16 °C) have shown that KPA addition positively influences wine quality by significantly reducing unstable protein fractions by up to 92% and improving protein stability . Neither KPA addition nor storage temperature adversely affected the volatile aroma profile, as confirmed by principal component analysis (PCA) [13].

As reported by Wang et al. [14], five dry red wines were analyzed to compare different tartrate stabilization approaches. Cold stabilization was found to effectively achieve tartrate stability but also significantly altered color and phenolic profiles, particularly when the treatment duration was extended. In contrast, tartrate-stabilizing agents containing KPA, GA, and sulfur dioxide, as well as those containing GA and sulfur dioxide, demonstrated superior performance by ensuring tartrate stability while better preserving color and phenolic compounds.

4.5 Gum arabic

Gum arabic is a complex polysaccharide obtained from the exudates of *Acacia senegal* and *Acacia seyal* trees. Structurally, GA is a highly branched polysaccharide composed of a galactose backbone linked to arabinose and galactose residues (arabinogalactan), with terminal residues typically consisting of rhamnose or glucuronic acid. Besides the glycan fraction, GA may contain a minor protein component known as arabinogalactan protein (AGP), which is characterized by serine and hydroxyproline-rich polypeptide chains [17].

Comparative studies indicate that GA generally exhibits lower stabilization efficiency than other tartrate-stabilizing additives [6, 15]. Guise et al. [6] reported that GA can enhance tartaric stability; however, its stabilizing effect was not consistent across all wine types and was strongly influenced by both the composition of the wine matrix and the degree of tartrate instability prior to treatment [6]. These findings suggest that GA is best suited for wines exhibiting low to moderate tartrate instability. Furthermore, evaluations of two distinct GA preparations revealed significant differences in performance, indicating that structural characteristics play a critical role in determining its effectiveness. GA interacts with tartrate molecules to form a polymeric coating that creates a hydrophilic layer around particles. This layer enhances the solubility of tartrates in wine, inhibits the formation and growth of crystal nuclei, and helps maintain the wine in a metastable state where crystals do not easily precipitate. Typically, it is added shortly before filtration and bottling. However, its use is not recommended for wines intended for aging or long-term storage, as it interferes with the natural formation of sediments that normally occur during long-term aging [59].

The action of GA is generally described as stabilizing rather than clarifying [58]. It functions as a protective colloid by adsorbing onto the surface of hydrophobic colloidal particles, thereby preventing their aggregation and subsequent precipitation. Uronic acid moieties have traditionally been considered responsible for electrostatic interactions with other wine components, such as potassium and calcium ions. Nevertheless, Nigen et al. [67] demonstrated that the protein fraction of AGP plays a dominant role in the electrostatic binding of Ca^{2+}, in contrast to the previously assumed primary contribution of uronic acids.

Despite its moderate standalone effectiveness, GA can be used in combination with other additives like CMC or KPA, where synergistic effects have been observed. This combined use is particularly valuable in red wines, as GA helps prevent color precipitation and improves mouthfeel, ensuring both partial tartaric stability and the long-term preservation of chromatic attributes without compromising overall stabilization performance [12, 14, 16].

5. Regulatory framework and legal limits

The use of enological additives in wine stabilization is strictly regulated to ensure product safety, technological necessity, and the preservation of wine authenticity. At the international level, the OIV establishes reference standards through the *International Enological Codex* and the *International Code of Enological Practices*. These standards are widely adopted or adapted within national and regional regulatory frameworks, including those of the European Union and Mercosur. Among the stabilizing agents most commonly used today are MTA, CMC, MPs, KPA, and GA.

Within the European Union (EU), Commission Delegated Regulation (EU) 2019/934 on authorized enological practices is fully aligned with OIV standards and establishes identical maximum limits for the use of tartaric acid stabilizers [68]. In the United States, authorization and conditions of use are defined by the Alcohol and Tobacco Tax and Trade Bureau (TTB), in coordination with the Food and Drug Administration (FDA). In Australia, regulation is overseen by Food Standards Australia New Zealand (FSANZ) under the Food Standards Code. **Table 1** summarizes the maximum permitted concentrations of tartrate stabilizers.

Metatartaric acid is authorized in the EU and Australia [69, 70], but is not permitted for use in the United States [71]. According to OIV practices, its application is limited to a maximum dose of 0.1 g/L [7, 45]. This restriction reflects its temporary effectiveness and susceptibility to hydrolysis over time, which may reduce tartrate stability and, if overdosed, potentially compromise wine clarity [59].

Additive	OIV	USA	Australia	Reference
Metatartaric acid	0.1	Not authorized	GMP*	[7, 45, 69, 70]
Carboxymethyl cellulose	0.2	GMP*	GMP*	[7, 45, 69, 70]
Mannoproteins	GMP*	0.4	0.4	[7, 45, 69, 70]
Potassium polyaspartate	0.1	0.1	0.1	[7, 45, 69, 70]
Gum arabic	Red liqueur wines: 0.8; other wines: 0.3	1.9	GMP*	[7, 45, 69, 70]

**For additives with no specific numerical maximum limit set, their use must be limited by GMP.*

Table 1.
Authorized maximum levels of tartaric stabilizers in wine (g/L).

For CMC, OIV regulations establish a maximum concentration of 0.2 g/L for white, rosé, and sparkling wines [7, 45]. This relatively low limit is related to its strong interactions with wine colloids and its potential to induce turbidity, particularly in protein-rich wines [71]. In the United States and Australia, its use must be limited by good manufacturing practice (GMP) [69, 70].

Mannoproteins occupy a distinct regulatory position. No numerical maximum limit is specified by the OIV [7, 45], as MPs are naturally occurring components of wine, and their residual concentrations can vary widely depending on winemaking practices. In contrast, both the United States and Australia impose a maximum addition level of 0.4 g/L of wine [69, 70].

KPA represents a more recent tartrate stabilizer. It is authorized by the OIV with a maximum dosage of 0.1 g/L [7, 45]. Exceeding this concentration may, in some cases, lead to increased wine turbidity. Notably, KPA is the only tartaric stabilizing agent for which there is broad consensus among international regulatory authorities regarding permitted usage levels [69].

Under OIV regulations, the maximum permitted dose of GA varies according to wine type. The limit is set at 0.8 g/L for red liqueur wines and 0.3 g/L for other wines [7, 45]. These differentiated limits reflect the higher phenolic content of liqueur wines, in which sensory quality is less affected by GA addition than in other wine styles. By contrast, the TTB permits its use up to 1.9 g/L [70], while in Australia, no specific maximum limit has been established [69].

Overall, the regulatory framework governing the use of enological additives reflects a careful balance among technological functionality, consumer protection, and respect for the natural composition and authenticity of wine.

6. Conclusions

Tartaric instability remains one of the most significant physicochemical challenges in winemaking, directly affecting wine limpidity, commercial quality, and consumer acceptance. The precipitation of KHT and CaT depends on pH, temperature, alcohol content, ionic composition, and natural colloids. Although supersaturation is a prerequisite for crystallization, the kinetics of nucleation and crystal growth are strongly modulated by wine matrix components such as proteins, polysaccharides, and phenolic compounds.

Analytical tools, ranging from traditional cold tests to rapid conductimetric and spectroscopic methods, support the evaluation of tartaric instability before bottling. At the same time, authorized additives, such as MTA, CMC, MPs, KPA, and GA, provide practical alternatives to cold stabilization, each with specific advantages and limitations related to their mechanism of action, wine compatibility, and regulatory requirements.

Ultimately, effective tartaric stabilization requires an integrated approach that combines scientific knowledge, accurate analytical evaluation, regulatory compliance, and consideration of the sensory and compositional integrity of the wine.

Acknowledgments

The authors used artificial intelligence (AI)-assisted language tools solely for English grammar and readability improvement. All scientific content, literature selection, interpretation, and conclusions were developed and critically reviewed by the authors, who retain full responsibility for the content of this chapter.

Conflict of Interest

The authors declare no conflict of interest.

Funding

This work was supported by FCT-Portugal, CQ-VR (UID/PRR2/00616/2025; https://doi.org/10.54499/UID/PRR2/00616/2025) and by Project Vine & Wine Portugal – Driving Sustainable Growth Through Smart Innovation, Application No. C644866286-00000011, co-financed within the scope of the Mobilizing Agendas for Business Innovation, under Reg. (EU) 2021/241, in the Plano de Recuperação e Resiliência (PRR) for Portugal, in its Component 5 – Capitalization and Business Innovation.

Author details

Fernanda Cosme[1,2]*, Juliana Milheiro[1], Luís Moreira[1], Sandrine Ferreira[1], Luís Filipe-Ribeiro[1], Mário Bezerra[1], Miguel Ribeiro[1,3] and Fernando M. Nunes[1,4]

1 Chemistry Research Center – Vila Real (CQ-VR), Food and Wine Chemistry Lab, University of Trás-os-Montes and Alto Douro, Vila Real, Portugal

2 Department of Biology and Environment, School of Life and Environmental Sciences, University of Trás-os-Montes and Alto Douro, Vila Real, Portugal

3 Department of Genetics and Biotechnology, School of Life and Environmental Sciences, University of Trás-os-Montes and Alto Douro, Vila Real, Portugal

4 Department of Chemistry, School of Life and Environmental Sciences, University of Trás-os-Montes and Alto Douro, Vila Real, Portugal

*Address all correspondence to: fcosme@utad.pt; fnunes@utad.pt; jmribeiro@utad.pt

References

[1] Waterhouse AL, Sacks GL, Jeffery DW. Understanding Wine Chemistry Chichester, United Kingdom. John Wiley & Sons; 2016

[2] Tan Q, Leng Y, Wang J, Huang C, Yuan Y. Correlation and prediction of the solubility of the racemic tartaric acid–ethanol–water system with the NRTL model. Journal of Molecular Liquids. 2016;**216**:476–483

[3] Lasanta C, Gómez J. Tartrate stabilization of wines. Trends in Food Science & Technology. 2012;**28**(1):52–59

[4] Boulton RB, Singleton VL, Bisson LF, Kunkee RE. Principles and practices of winemaking. New York, USA: Chapman & Hall; 1996

[5] Cosme F, Filipe-Ribeiro L, Coixão A, Bezerra M, Nunes FM. Efficiency of alginic acid, sodium carboxymethylcellulose, and potassium polyaspartate as calcium tartrate stabilizers in wines. Foods. 2024;**13** (12):1880

[6] Guise R, Filipe-Ribeiro L, Nascimento D, Bessa O, Nunes FM, Cosme F. Comparison between different types of carboxymethylcellulose and other oenological additives used for white wine tartaric stabilization. Food Chemistry. 2014;**156**:250–257

[7] Organisation Internationale de la Vigne et du Vin (OIV). International Oenological Codex. Official ed. Dijon (France): OIV; 2025

[8] Sprenger S, Hirn S, Dietrich H, Will F. Metatartaric acid: Physicochemical characterization and analytical detection in wines and grape juices. European Food Research and Technology. 2015;**241**(6):785–791. DOI: 10.1007/s00217-015-2503-1

[9] Cosme F, Oliveira R, Filipe-Ribeiro L, Nunes FM. Wine volatilome as affected by tartaric stabilization treatments: Cold stabilization, carboxymethylcellulose and metatartaric acid. Foods. 2024;**13** (17):2734. DOI: 10.3390/foods13172734

[10] Filipe-Ribeiro L, Milheiro J, Guise R, Vilamarim R, Fraga JB, Martins-Gomes C, Nunes FM, Cosme F. Efficiency of carboxymethylcellulose in red wine tartaric stability: Effect on wine phenolic composition, chromatic characteristics and colouring matter stability. Food Chemistry. 2021;**360**:129996

[11] Pittari E, Catarino S, Andrade MC, Ricardo-da-Silva JM. Preliminary results on tartaric stabilization of red wine by adding different carboxymethylcelluloses. Ciência e Técnica Vitivinícola. 2018;**33**:47–57

[12] Bosso A, Motta S, Panero L, Lucini S, Guaita M. Use of potassium polyaspartate for stabilization of potassium bitartrate in wines: Influence on colloidal stability and interactions with other additives and enological practices. Journal of Food Science. 2020;**85**(8):2406–2415. DOI: 10.1111/1750-3841.15342

[13] Natolino A, Tat L, Gallo A, Roman T, Celotti E. Use of potassium polyaspartate on white wines: Interaction with proteins and aroma compounds. Food Research International. 2023;112768. DOI: 10.1016/j.foodres.2023.112768

[14] Wang H, Fan X, Zhou Y, Li D, Ling M, Wei X, Liu P, Zhang X. Balancing wine stability with color and phenolic quality: Insights into the impact of fining agents and tartrate stabilization methods. Food Chemistry. 2025;**493**(1):145731. DOI: 10.1016/j.foodchem.2025.145731

[15] Bosso A, Salmaso D, De Faveri E, Guaita M, Franceschi D. The use of carboxymethylcellulose for the tartaric stabilization of white wines, in comparison with other oenological additives. Vitis. 2010;**49**(2):95–99

[16] Martínez-Pérez MP, Bautista-Ortín AB, Durant V, Gómez-Plaza E. Evaluating alternatives to cold stabilization in wineries: The use of carboxymethyl cellulose, potassium polyaspartate, electrodialysis and ion exchange resins. Foods. 2020;**9**(9):1275

[17] Nie SP, Wang C, Cui SW, Wang Q, Xie MY, Phillips GO. A further amendment to the classical core structure of gum arabic (Acacia Senegal). Food Hydrocolloids. 2013;**31**(1):42–48

[18] Hernandez LM, Ballou L, Alvarado E, Gillece-Castro BL, Burlingame AL, Ballou CE. A new Saccharomyces cerevisiae mnn mutant N-linked oligosaccharide structure. Journal of Biological Chemistry. 1989;**264**(20):11849–11856

[19] Varki A, Cummings RD, Aebi M, Packer NH, Seeberger PH, Esko JD, Kornfeld S. Symbol nomenclature for graphical representations of glycans. Glycobiology. 2015;**25**(12):1323–1324

[20] Berg HW, Keefer R. Analytical determination of tartrate stability in wine. I. Potassium bitartrate. American Journal of Enology and Viticulture. 1958;**9**(4):180–193. DOI: 10.5344/ajev.1958.9.4.180

[21] Barlow DA, Gregus J. The kinetics of homogeneous and two-step nucleation during protein crystal growth from solution. International Journal of Chemical Kinetics. 2019;**51**(11):840–847. DOI: 10.1002/kin.2131

[22] Dunsford P, Boulton RB. The kinetics of potassium bitartrate crystallization from table wines. II. Effect of particle size, particle surface area and agitation. American Journal of Enology and Viticulture. 1981;**32**(2):100–105. DOI: 10.5344/ajev.1981.32.2.100

[23] Ribéreau-Gayon P, Glories Y, Maujean A, Dubourdieu D. Handbook of Enology, the Chemistry of Wine: Stabilization and Treatments: Second Edition. Vol. England: John Wiley & Sons; 2nd ed. **2** 2006. DOI: 10.1002/0470010398

[24] Flanzy C. Œnologie: Fondements Scientifiques Et Technologiques. 2nd ed. Mundi Prensa Libros: A. M. V. Éditions; 2003; p. 1310

[25] Gonçalves F, Fernandes C, Cameira Dos Santos P, De Pinho MN. Wine tartaric stabilization by electrodialysis and its assessment by the saturation temperature. Journal of Food Engineering. 2003;**59**(2–3):229–235. DOI: 10.1016/S0260-8774(02)00462-4

[26] Wang S. Mechanism of tartar formation in wine and methods to tartar stabilization. Liquor Making. 2004;**05**:60–61

[27] Abguéguen O, Boulton RB. The crystallization kinetics of calcium tartrate from model solutions and wines. American Journal of Enology and

Viticulture. 1993;**44**(1):65–73. DOI: 10.5344/ajev.1993.44.1.65

[28] Berovic M, Kosmerl T. Monitoring of potassium hydrogen tartrate stabilization by conductivity measurement. Acta Chim Slov. 2008;**55**:535–540

[29] Lambri M, Colangelo D, Dordoni R, De Faveri DM. The effects of different protein: tannin ratios on the tartrate-holding capacity of wine model solutions. Food Research International. 2014;**62**:441–447. DOI: 10.1016/j.foodres.2014.03.044

[30] McKinnon AJ, Williams PJ, Scollary GR. Influence of uronic acids on the spontaneous precipitation of calcium L-(+)-tartrate in a model wine solution. Journal of Agricultural and Food Chemistry. 1996;**44**(6):1382–1386. DOI: 10.1021/jf950111v

[31] Pellerin P, Doco T, Scollary GR. The influence of wine polymers on the spontaneous precipitation of calcium tartrate in a model wine solution. International Journal of Food Science and Technology. 2013;**48**(12):2676–2682. DOI: 10.1111/ijfs.12264

[32] McKinnon AJ, Scollary GR, Solomon DH, Williams PJ. The influence of wine components on the spontaneous precipitation of calcium L (+)-tartrate in a model wine solution. American Journal of Enology and Viticulture. 1995;**46**(4):509–517

[33] Postel W. La solubilité et la cinétique de cristalisation du tartrate de calcium dans le vin. La Journée Vinicole. 1984;**16688**:2–10

[34] Maujean A, Sausy L, Vallée D. Détermination de la sursaturation en bitartrate de potassium d'un vin: Quantification des effets colloïdes-protecteurs. Rev Fr Œnol. 1985;**100**:39–50

[35] Moutounet M, Saint-Pierre B, Battle JL, Escudier JL. Stabilisation tartrique: Détermination du degré d'instabilité des vins. Mesure de l'efficacité des inhibiteurs de cristallisation. In: Proceedings of the VIth International Oenology Symposium; 1999;531–534

[36] Soares P, Geraldes V, Fernandes C, Dos Santos PC, de Pinho MN. Wine tartaric stabilization by electrodialysis: Prediction of required deionization degree. American Journal of Enology and Viticulture. 2009;**60**(2):183–188. DOI: 10.5344/ajev.2009.60.2.183

[37] Muller-Späth H. Un nouveau procédé de prévention des précipitations de bitartrate de potassium: Le procédé contact. Annales de Technologie Agricole. 1977;**27**(1):333–337

[38] Moutounet M, Saint-Pierre B, Battle JL, Escudier JL. Le stabilisateur tartrique: Principe et description du procédé. Rev Fr Œnol. 1997;**162**:15–17

[39] Moutounet M, Bouisson D, Escudier JL. Détermination du degré d'instabilité tartrique (DIT): Principes et applications. Rev Fr Œnol, Cah Tech. 2010;**242**:24–28

[40] Bosso A, Motta S, Petrozziello M, Guaita M, Asproudi A, Panero L. Validation of a rapid conductimetric test for the measurement of wine tartaric stability. Food Chemistry. 2016;**212**:821–827. DOI: 10.1016/j.foodchem.2016.06.044

[41] Boulangé-Petermann L, Vernhet A, Dupre K, Moutounet M. KHT cold

stabilization: A scanning electron microscopy study of the formation of surface deposits on stainless steel in model wines. VITIS. 2015;**38** (1):43

[42] Malacarne M, Bergamo L, Bertoldi D, Nicolini G, Larcher R. Use of Fourier transform infrared spectroscopy to create models forecasting the tartaric stability of wines. Talanta. 2013;**117**:505–510. DOI: 10.1016/j.talanta.2013.08.036

[43] Versari A, Laghi L, Thorngate JH, Boulton RB. Prediction of colloidal stability in white wines using infrared spectroscopy. Journal of Food Engineering. 2011;**104**(2):239–245. DOI: 10.1016/j.jfoodeng.2010.12.015

[44] Ding H, Hou R, Li Y, Zhang B, Zhao B, Liu K. Effect of different carboxymethyl cellulose structure parameters on tartrates stability of red wine: Viscosity and degree of substitution. Food Additives & Contaminants: Part A. 2020;**37**: 1099–1109

[45] Organisation Internationale de la Vigne et du Vin (OIV). International Code of Oenological Practices. Dijon (France): OIV; 2025

[46] Bosso A, Panero L, Petrozziello M, Asproudi A, Motta S, Guaita M. Use of polyaspartate as inhibitor of tartaric precipitations in wines. Food Chemistry. 2015;**185**:1–6. DOI: 10.1016/j.foodchem.2015.03.099

[47] Crachereau JC, Gabas N, Blouin J, Hébrard B, Maujean A. Stabilisation tartrique des vins par la carboxyméthylcellulose. Bulletin de l'OIV. 2001;**74**:841–842

[48] Rodriguez-Clemente R, Correa-Gorospe I, J.J DC. A new method for the stabilization of wines with respect to potassium bitartrate precipitation. American Journal of Enology and Viticulture. 1990;**41**:16–20

[49] Greeff A, Robillard B, Toit W. Short- and long-term efficiency of carboxymethylcellulose (CMC) to prevent crystal formation in South African wine. Food Additives & Contaminants: Part A. 2012;**29**: 1374–1385

[50] Lubbers S, Léger B, Charpentier C, Feuillat M. Effet colloïde-protecteur d'extraits de parois de levure sur la stabilité tartrique d'une solution hydro-alcoolique modèle. Journal International Des Sciences de la Vigne Et du Vin. 1993;**27**:13–22

[51] Wucherpfennig K, Dietrich H, Götz W, Rötz S. Influence of colloids on tartrate crystallization with special consideration of wine stabilization by carboxymethylcellulose. Die Weinwirtschaft-Technik. 1984;**1**:13–23

[52] Wucherpfennig K, Otto K, Kern U. Practical application of carboxymethylcellulose. Die Weinwirtschaft-Technik. 1988;**5**:13–19

[53] Gerbaud V, Gabas N, Blouin J, Crachereau JC. Study of wine tartaric acid salt stabilization by addition of carboxymethylcellulose (CMC): Comparison with the "protective colloids" effect. Journal International Des Sciences de la Vigne Et du Vin. 2010;**44**:231–242

[54] Bowyer P, Gouty C, Moine V, Marsh R, Battaglene T. CMC: A new potassium bitartrate stabilisation tool. Australian & New Zealand

Grapegrower & Winemaker. 2010;**558**:65–68

[55] Salagoïty MH, Guyon F, René L, Gaillard L, Lagrèze C, Domec A, Baudouin M, Médina B. Quantification method and organoleptic impact of added carboxymethylcellulose in dry white wine. Food & Function. 2011;**3**:380–384

[56] Moutounet M, Bouissou D, Escudier JL. Effets de traitement de stabilization tartrique de vins rouges par une gomme de cellulose (carboxymethylcellulose). Infowine. 2010;**6**:11–19

[57] Claus H, Tenzer S, Sobe M, Schlander M, König H, Fröhlich J. Effect of carboxymethyl cellulose on tartrate salt, protein and colour stability of red wine. Australian Journal of Grape and Wine Research. 2014;**20**:186–193

[58] Sommer S, Dickescheid C, Harbertson JF, Fischer U, S.d C. Rationale for haze formation after carboxymethyl cellulose (CMC) addition to red wine. Journal of Agricultural and Food Chemistry. 2016;**64**:6879–6887

[59] Cui W, Wang X, Han S, Guo W, Meng N, Li J, Sun B, Zhang X. Research progress of tartaric acid stabilization on wine characteristics. Food Chemistry: X. 2024;**23**:101728. DOI: 10.1016/j.fochx.2024.101728

[60] Pascotto K, Leriche C, Caillé S, Violleau F, Boulet J-C, Geffroy O, et al. Study of the relationship between red wine colloidal fraction and astringency by asymmetrical flow field-flow fractionation coupled with multi-detection. Food Chemistry. 2021;**361**:130104

[61] Snyman C, Mekoue Nguela J, Sieczkowski N, Marangon M, Divol B. Impact of mannoproteins from different yeast species on wine properties. OENO One. 2023;**57**(4). DOI: 10.20870/oeno-one.2023.57.4.7711

[62] Rodrigues A, Ricardo-da-silva JM, Lucas C, Laureano O. Influence of fining and tartaric stabilisation procedures on white wine mannoprotein content. South African Journal of Enology and Viticulture. 2012;**33**(1):88–94

[63] Snyman C, Mekoue Nguela J, Sieczkowski N, Divol B, Marangon M. Characterization of mannoprotein structural diversity in wine yeast species. Journal of Agricultural and Food Chemistry. 2023;**71**(49):19727–19738

[64] Lubbers S, Voilley A, Feuillat M, Charpentier C. Influence of Mannaproteins from Yeast on the Aroma Intensity of a Model Wine. LWT-Food Science and Technology. 1994;**27**(2):108–114

[65] Silva-Barbieri D, Salazar FN, López F, Brossard N, Escalona N, Pérez-Correa JR. Advances in white wine protein stabilization technologies. Molecules. 2022;**27**(4):1251

[66] Ribeiro T, Fernandes C, Nunes FM, Filipe-Ribeiro L, Cosme F. Influence of the structural features of commercial mannoproteins in white wine protein stabilization and chemical and sensory properties. Food Chemistry. 2014;**159**:47–54

[67] Nigen M, Valiente RA, Iturmendi N, Williams P, Doco T, Moine V, Sanchez C. The colloidal stabilization of young red wine by Acacia Senegal gum: The involvement of the protein backbone from the

protein-rich arabinogalactan-proteins. Food Hydrocolloids. 2019;**97**:105176

[68] European Commission. Commission Delegated Regulation (EU) 2019/934. Official Journal of the European Union, L 149. 2019;7:1–52

[69] Alcohol and Tobacco Tax and Trade Bureau (US). Code of Federal Regulations. Title 27, Chapter I, Subchapter A, Part 24—Wine. Washington (DC) Government Publishing Office; 2025. https://www.ecfr.gov/ [Accessed: 2026-January-30]

[70] Australia New Zealand Food Standards Code: Schedule 15 – Substances that may be used as food additives [Internet] Federal Register of Legislation. 2020. Available from: https://www.legislation.gov.au/Details/F2020C00203 [Accessed: 2026-January-22]

[71] Motta S, Bollito A, Guaita M, Petrozziello M, Panero L, Bosso A. Impiego della carbossimetilcellulosa e uso di altri additivi nella stabilizzazione tartarica dei vini rossi. Bulletin de l'OIV. 2009;**82** (941–943):367–377

Chapter 5

Smart Packaging Additives: Bridging Post-Harvest Loss Control and Consumer Appeal

Ravindra Naik, Ramabhau Tumadu Patil, Aswathy Thankappan, Anjana Kallampilly and Gowtham Manoharan

Abstract

Food packaging systems are designed to preserve freshness and structural integrity. Recent advancements have led to the development of novel packaging systems collectively referred to as smart packaging technology. This chapter explores the growing role of smart packaging additives in minimizing post-harvest losses and meeting modern consumer expectations. It highlights two key types: active additives (e.g., oxygen scavengers, ethylene absorbers, and antimicrobial agents), which preserve product quality, and intelligent additives (e.g., time–temperature indicators, freshness sensors, and radio frequency identification [RFID] tags), which provide real-time information on food safety and handling. It also discusses their applications across fruits, vegetables, dairy, meat, and ready-to-eat products, emphasizing how they extend shelf life, improve traceability, and enhance consumer trust. Furthermore, the market drivers, like the clean-label movement and sustainability, are discussed, along with challenges related to cost, regulation, and waste management. Overall, it provides a comprehensive overview of how these innovations can bridge the gap between industrial preservation needs and the rising demand for freshness, transparency, and safety in food systems.

Keywords: active packaging, consumer demand, food safety, intelligent packaging, post-harvest loss, smart packaging, supply chain, sustainability

1. Introduction

Post-harvest losses, especially in perishable commodities like fruits, vegetables, dairy, and meat, are a significant concern globally. Traditional preservation methods are no longer sufficient to meet modern consumer demands for freshness, minimal processing, clean labels, and traceability. Fruits and vegetables have the highest share of food wastage at around 45% [1], which highlights the urgent need to reduce the generation of food waste and develop multifunctional packaging systems. Recent advancements in food packaging systems have paved the way for creative packaging techniques known as

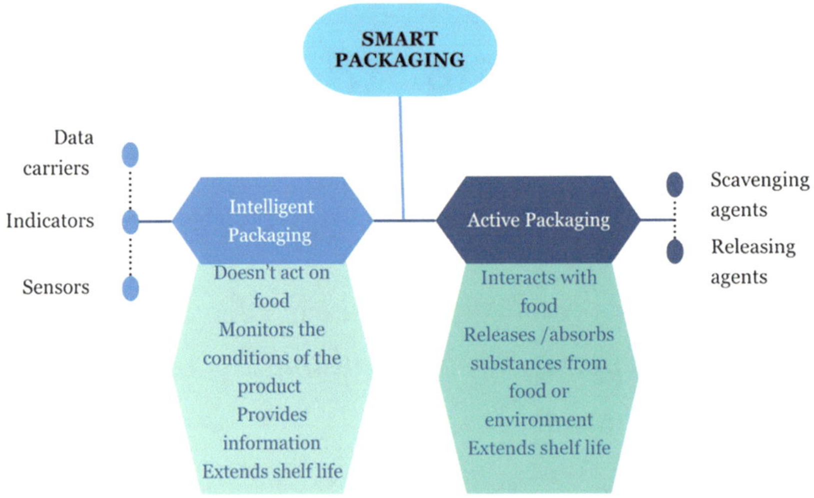

Figure 1.
Classification of smart packaging systems [2].

"smart packaging technology," which include mechanisms to accurately track the alterations occurring within the food product [2]. This chapter discusses in detail both the active and intelligent packaging additives and their principles. The classification of smart packaging systems is shown in **Figure 1**. In recent years, smart packaging additives like oxygen scavengers, moisture regulators, ethylene absorbers, and antimicrobial agents have emerged as an innovative option to enhance preservation efficiency and extend shelf life [3, 4] and are highlighted in the subsections of this chapter. These additives actively interact with the packaged product or environment, helping to inhibit microbial growth, delay ripening, and maintain optimal physicochemical conditions.

However, intelligent packaging transforms traditional containers into responsive systems that monitor, detect, and communicate changes in product quality or environmental conditions [5], thus promising a solution that extends shelf life, reduces spoilage, and enhances consumer trust. The subsequent sections explore how intelligent packaging additives, such as time–temperature indicators (TTIs), freshness indicators, and radio frequency identification (RFID)/near-field communication (NFC) tags with embedded sensors, address specific post-harvest challenges.

Consumer preferences increasingly drive food packaging innovations, emphasizing clean-label products, transparency, convenience, and sustainability [6]. The following subsections examine the significance of these key market drivers within the food sector. These factors shape the development of smart and functional packaging solutions that align with modern market demands.

2. Active packaging additives

2.1 Oxygen scavengers

The presence of oxygen inside food packaging initiates oxidative reactions and promotes the growth of specific microorganisms. This can lead to spoilage or quality degradation, causing foul odors, undesirable color changes, off-flavor formation, loss of nutritional value, and other defects. The use of oxygen scavengers can prevent these issues [2]. Oxygen scavengers are materials embedded within food packaging that actively remove residual oxygen from the surrounding environment. It is an active packaging approach designed to remove dissolved oxygen or oxygen present in the headspace, thereby helping to enhance the shelf life of the product. Oxygen scavengers function by chemically or enzymatically reacting with oxygen, thereby preserving the quality and shelf life of packaged foods – particularly in high-fat or protein-rich foods like meat, cheese, and nuts [7]. Some of the common oxygen scavengers are discussed below.

2.1.1 Iron powder sachets

This is the most commonly used oxygen scavenger in the food industry, typically incorporated as pads or sachets containing metallic reducing agents like powdered iron oxide, ferrous carbonate, or metallic platinum. These systems depend on moisture to initiate oxygen absorption via iron oxidation, thereby limiting their effectiveness in dry food products [8].

2.1.2 Ascorbic acid systems

Ascorbic acid is a six-carbon compound. It works on the principle of ascorbate oxidation to dehydroascorbic acid, which is relatively harmless. Metal ions, like iron, accelerate the oxidation of ascorbic acid, converting it into dehydroascorbic acid [9].

2.1.3 Enzymatic solutions

Enzymes are utilized to consume oxygen, thereby reducing or eliminating it from the environment and protecting sensitive products from oxidation. The most commonly used enzymatic oxygen scavenging systems involve glucose oxidase (GOD). It is used to remove oxygen by converting glucose to gluconic acid and water. Other commonly used enzymes include laccase and ethanol oxidase [10].

These systems are especially useful in vacuum-sealed or modified atmosphere packaging (MAP), where even minimal oxygen can accelerate spoilage. By actively binding and removing oxygen molecules from the package, these scavengers help maintain the product's freshness and nutritional quality, thereby extending its shelf life. They are used for the preservation of all types of packaged foods, foods containing oils or fats, rice, bakery products, juice, meat, milk products, pharmaceuticals, and fruits and vegetables [10, 11]. They have enhanced the storage life of fish by effectively controlling fat oxidation and inhibiting the growth of microorganisms [8].

2.2 Moisture regulators

Moisture regulation is essential to protect both high-moisture and dry food products from the accumulation of excess moisture, which can lead to quality degradation of the food as well as deterioration of the packaging. Moisture accumulation inside food packaging can arise from several sources, including natural respiration of fresh produce, fresh meat, fish, poultry, condensation or sweating, and accumulation of humidity due to the metabolism of biomolecules [12]. The application of moisture regulators can remove accumulated moisture and maintain the quality of both food products and packaging. Additionally, it lowers water activity and reduces microbial deterioration in high-water-activity foods such as fruits, vegetables, and meats. For this reason, moisture regulators are incorporated into active packaging technology. This aids in controlling the moisture levels inside the package by either absorbing or releasing excess water, thereby maintaining optimal relative humidity [13].

Packaging that incorporates desiccants such as silica gel or calcium chloride helps maintain optimal relative humidity within the package. Silica gel is among the most widely used desiccants in food packaging due to its high moisture-adsorbing capacity, non-toxic nature, chemical stability, and high relative humidity tolerance [14]. It is a porous, granular form of silicon dioxide (SiO_2). The large surface area and presence of millions of tiny pores enhance the moisture absorption capacity of silica gel. Depending on the relative humidity within the packaging, silica gel can adsorb up to 40% of its weight in moisture [15]. Calcium chloride can absorb a significant amount of water from the surrounding environment relative to its weight. It absorbs moisture through deliquescence (a process by which a solid material absorbs sufficient moisture from the atmosphere to dissolve into a liquid solution), which forms liquid brine once it absorbs enough water [16]. These are particularly important for bakery products, snack foods, and powdered items, where excess moisture can lead to microbial growth, clumping, or textural loss. A comparison between silica gel and calcium chloride is given in **Table 1**.

In contrast, humectant-based additives may be used to retain moisture in products like fresh produce, preventing dehydration and wilting during extended shelf life. They function by attracting and binding water molecules and by preventing food from losing water. Being hygroscopic, they absorb moisture from the surrounding environment or the food itself, thereby preventing it from drying out [18, 19]. Furthermore, by lowering water activity, they help suppress the growth of microorganisms and extend the shelf life of food products. Humectant-based additives maintain the desirable qualities of food, like softness and pliability. The commonly used humectants include glycerol, sorbitol, propylene glycol, sugars like sucrose, honey, salt, fruit juices, and certain polysaccharides [19–21] (see **Figure 2**).

Food products utilizing humectants include baked goods, confectionery, meat products, dried fruits, cereals, dairy products, and snacks [18, 22, 23]. Humectants help maintain freshness and softness in baked goods and dried fruits by preventing excessive drying. In meat products, they can be used to improve moisture retention and texture [20, 24, 25]. They help maintain crispiness in cereals and snacks, and they also prevent candies and chocolates from becoming hard [22].

Properties	Silica gel	Calcium chloride
Moisture capacity	Moderate	Very high
Type of reaction	Adsorptive	Deliquescent
Speed of moisture absorption	Slower	Faster
Reusability	Reusable after drying	Not reusable

Source: Study by Flórez-Méndez and López [17].

Table 1.
Comparison between silica gel and calcium chloride.

Figure 2.
Humectants in food packaging [19–22].

2.3 Ethylene absorbers

Storage of climacteric fruits is difficult because they continue to ripen even after harvesting due to ethylene production. Ethylene (C_2H_4) is a natural plant hormone that increases the respiration rate and accelerates ripening and senescence in horticultural products such as fruits, vegetables, and flowers [26]. Ethylene is beneficial in many cases, such as inducing flowering, enhancing color development, stimulating root growth, contributing bitter flavor, aiding seed germination, and supporting stress response, but in many horticultural situations, it is necessary to remove ethylene or suppress its negative effects. Ethylene absorbers in active packaging regulate ethylene levels, thereby prolonging the shelf life of fresh produce [27].

Ethylene absorbers incorporated into packaging are available in various forms, such as sachets placed inside the package, coatings on packaging material, or as ethylene-scavenging active films [3]. Commonly used ethylene absorbers are in the form of sachets containing potassium permanganate ($KMnO_4$), activated carbon, or zeolites. This helps adsorb or neutralize ethylene, thus slowing the ripening process [15].

Ethylene absorbers function through various mechanisms as shown in **Figure 3**.

2.3.1 Potassium permanganate

It is a widely used chemical agent in food packaging to absorb and oxidize ethylene gas. It chemically reacts with ethylene and converts ethylene into carbon

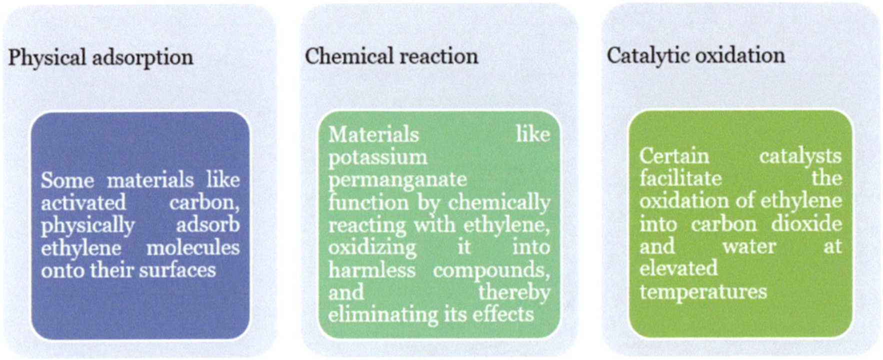

Figure 3.
Mechanisms of ethylene absorbers [28].

dioxide (CO_2) and water (H_2O) by the oxidation process [28]. For bananas, $KMnO_4$ is the most commonly used ethylene absorber [27].

2.3.2 Activated carbon

Due to its high surface area and porous structure, it can trap and hold ethylene molecules, thus preventing the ripening process [29]. Activated carbon has been widely used to lower ethylene concentrations and preserve the quality of various fruits and vegetables during storage, including tomatoes and broccoli [30].

2.3.3 Zeolites

It is a highly porous natural mining material. Its adsorption mechanism involves both physisorption and chemisorption (a type of adsorption where the surface of one material (adsorbent) and another substance (adsorbate) form strong chemical bonds, like covalent or ionic), with exchanged zeolites exhibiting enhanced chemisorption for improved trace ethylene removal [28].

The mechanism of ethylene absorbers is depicted in **Figure 4**. Ethylene absorbers are especially useful for climacteric fruits like bananas, apples, kiwifruits, apricots, cucumbers, avocados, tomatoes, and mangoes, which continue to ripen after harvest, and vegetables such as carrots, potatoes, and asparagus [31]. Such additives are crucial in long-distance transport and export packaging, where ethylene build-up can lead to over-ripening and significant post-harvest losses.

2.4 Antimicrobial agents

Antimicrobial packaging involves the inclusion of substances that inhibit or destroy pathogenic and spoilage microorganisms. Antimicrobials prolong the lag phase of microbes or inactivate them, thereby slowing their growth rate and reducing their population. These may include naturally derived antimicrobials like nisin, lysozyme, chitosan, or essential oils (e.g., oregano, thyme), as well as silver nanoparticles (Ag NPs) or organic acids [32].

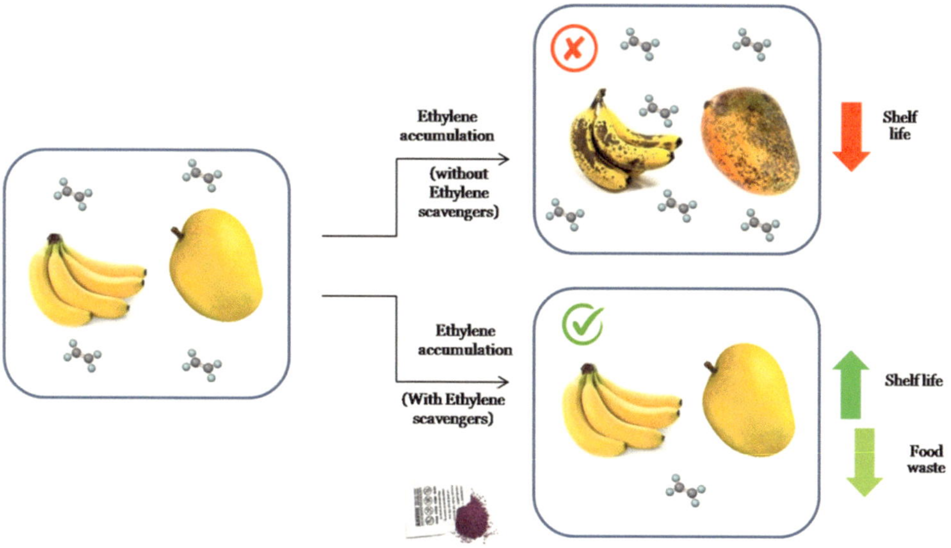

Figure 4.
Mechanism of ethylene absorbers (Adapted from the study by Gaikwad and Negi [27]).

2.4.1 Nisin

Nisin is effective in controlling Gram-positive bacteria, including some foodborne pathogens. It can be incorporated into multiple packaging materials, including edible films, to serve as a barrier against microbial contamination [33].

2.4.2 Lysozyme

It is an enzyme obtained from egg white that breaks down the peptidoglycan layer in the cell walls of Gram-positive bacteria, causing cell lysis and ultimately bacterial death [15].

2.4.3 Chitosan

This is a natural polymer derived from chitin, which is a cationic polymer with antimicrobial activity. It can be applied as a coating material to protect fresh produce from fungal degradation, thereby enhancing its shelf life [34].

Chicken drumsticks wrapped in multilayer low-density polyethylene (LDPE) active film containing chitosan were effective in inhibiting total aerobic mesophilic bacteria, coliforms, molds, and yeasts compared with the normal LDPE–Polyamide–LDPE bags [35].

2.4.4 Essential oils

They are a good source of volatile terpenoids and phenolic compounds, which give them strong activity against a wide spectrum of microorganisms. Essential oils can be directly incorporated into packaging materials or used in combination with encapsulation techniques to develop active packaging systems. Examples include oregano oil and thyme [34].

2.4.5 Silver nanoparticles

Ag NPs demonstrate broad-spectrum antimicrobial activity, acting against various pathogens, including bacteria, fungi, and certain viruses. Ag NPs have the ability to disrupt microbial cell membranes, interfere with DNA replication, inhibit enzyme activity, and can be incorporated into different packaging materials [15].

2.4.6 Organic acids

Organic acids reduce the pH of foods and suppress the growth of unwanted microorganisms. They occur in both dissociated and undissociated forms, but only the undissociated molecules can penetrate microbial cell walls, where they disrupt metabolism and exert their inhibitory action. Acids like lactic acid, sorbic acid, benzoic acid, and propionic acid can be used in food packaging as antimicrobial agents [12]. Sorbic acid is among the least harmful antimicrobial agents used in food products. The shelf life of bread and cheese wrapped with LDPE films containing benzoic acid can be significantly improved [34].

Antimicrobial agents are widely applied in the packaging of meat, poultry, seafood, and fresh-cut produce, where microbial spoilage poses a major safety and quality challenge. Among food packaging technologies, antimicrobial packaging has gained significant attention due to the wide range of materials used, the various application methods available, and its suitability for protecting diverse food products [36].

3. Intelligent packaging additives

In contrast to active packaging discussed in the previous section, intelligent packaging does not interact with the product or package; instead, it transforms traditional containers into responsive systems that monitor, detect, and communicate changes in product quality or environmental conditions [5]. Intelligent packaging additives are components incorporated into packages that facilitate the package in monitoring, detecting, or recording changes in the condition of a product or its environment, such as quality, freshness, temperature, humidity, or spoilage [37]. These additives provide the packaging with intelligent functionalities, transforming it from a passive barrier into an informative system, without necessarily interacting directly to preserve the product [38]. They do not directly preserve the product but provide real-time information that enhances safety, traceability, and consumer confidence.

3.1 Time–temperature indicators

TTIs are smart labels that visually reflect the cumulative exposure of a product to temperatures above its safe threshold. These indicators typically change color based on a chemical or enzymatic reaction that correlates with time and temperature exposure. A TTI may be defined as a device that integrates the thermal exposure over

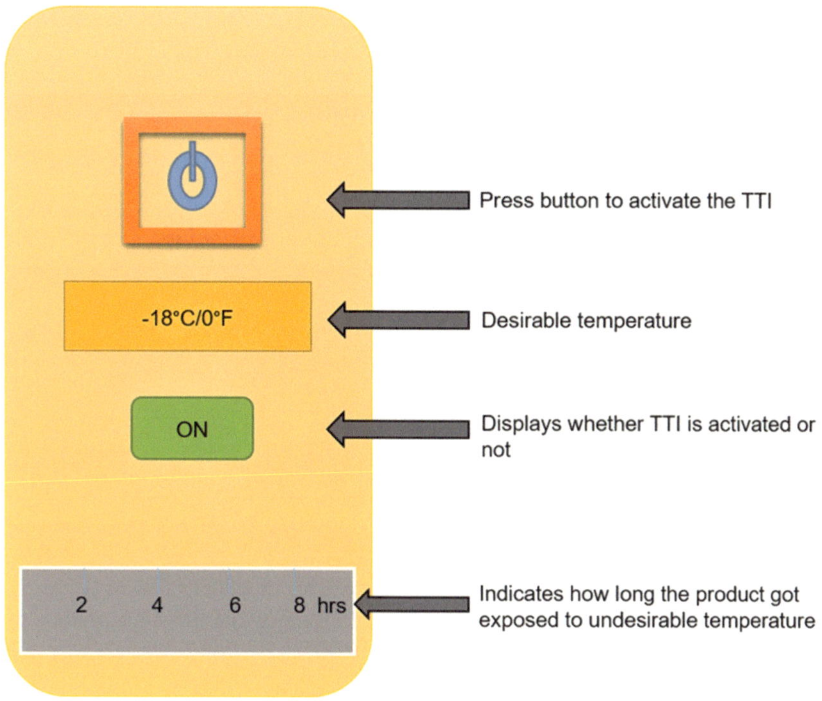

Figure 5.
Time–temperature indicator (adapted from the study by Vasuki et al. [2]).

time, accumulating the effects of such exposures and displaying a visible change, typically a color shift or other irreversible physical alteration, to indicate the product's thermal history [39]. **Figure 5** shows a schematic representation of TTI. TTIs are especially valuable for frozen and chilled food products such as seafood, dairy, and vaccines, where any deviation in the cold chain can compromise product integrity.

The working of a TTI is as described below [40, 41]:

(1) **Activation**: Some TTIs need an initial activation step, such as a physical trigger or exposure to UV light, to start working. Such TTIs begin time–temperature monitoring and recording of temperature changes only after activation.

(2) **Temperature sensing**: There are different types of TTIs.

- Diffusion-based: Heat causes a dye to migrate along a strip; the distance correlates with the exposure time and temperature.

- Polymer-based: Chemical reactions cause a visible color change due to polymerization.

- Enzyme-based: Enzymatic degradation alters pH, triggering a color shift.

- Microbial-based: Microbial growth leads to dye changes, indicating spoilage.

(3) **Color change**: When there is temperature abuse over time, an irreversible color change occurs. The extent of the color change corresponds to the cumulative time–temperature exposure.

(4) **Indication**: The final color of the TTI indicates the accumulated time–temperature history of the product. A full color change indicates spoilage of the product due to temperature abuse for a crucial duration. A partial color change indicates that the product was subjected to unfavorable temperature but can still be consumed, as that exposure was within the acceptable limit in terms of quality and safety.

These indicators are especially valuable in cold chain logistics for seafood, dairy, and vaccines. Some TTIs can even record the temperature history, helping identify where in the supply chain a breach occurred [42].

3.2 Freshness indicators

Freshness indicators detect spoilage-related changes, such as gas emissions (e.g., ammonia, hydrogen sulfide) or pH shifts [43]. They provide a visual cue, often a color change, that reflects the actual condition of the food rather than relying solely on expiry dates.

A colorimetric strip used in fish packaging changes from green to red as spoilage progresses due to amine release, as illustrated in **Figure 6**.

In meat or dairy, sensors may respond to microbial metabolites or changes in acidity.

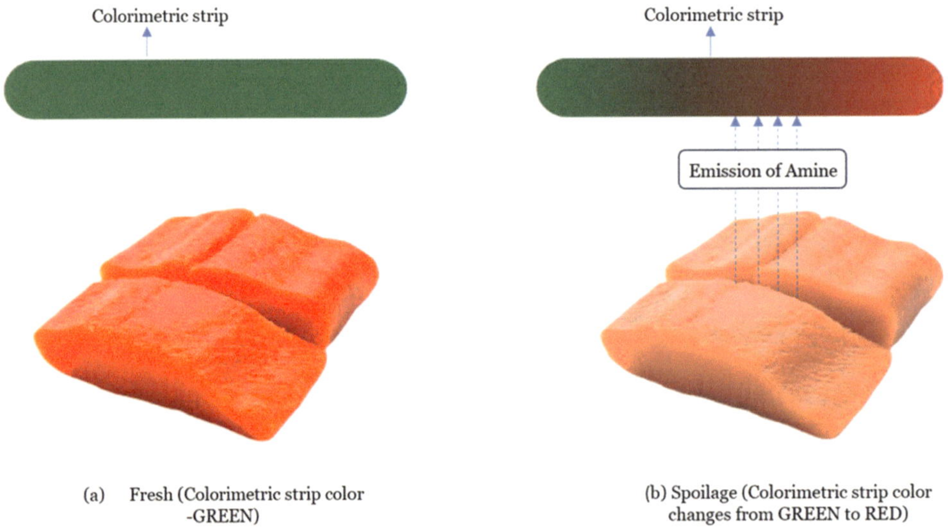

Figure 6.
Colorimetric strip in fish packaging (adapted from the study by Azeredo and Correa [44]).

These indicators equip consumers with the information needed to make informed decisions, thereby reducing unnecessary food wastage.

3.3 RFID/NFC tags with embedded sensors

RFID and NFC technologies facilitate wireless communication of real-time product data, including temperature, humidity, and movement during transport. RFID tags use electromagnetic fields of radio frequency to store and communicate product information in real time for their identification and traceability [45, 46]. **Figure 7** shows a schematic diagram depicting the working principle of RFID.

The working of RFID is as follows:

(1) **Initiation**: The reader emits a signal to activate the tag.

(2) **Activation**: After receiving a radio signal, the RFID tag gets activated.

(3) **Tag response and data transmission**: The tag transmits stored data, and the information is sent to the reader.

(4) **Processing**: Data is analyzed for action.

(5) **Automation**: Triggers alerts or system responses (e.g., alarms, door controls).

These technologies support traceability, authenticity, and real-time monitoring, which are especially useful for high-value perishables and export commodities [45, 46].

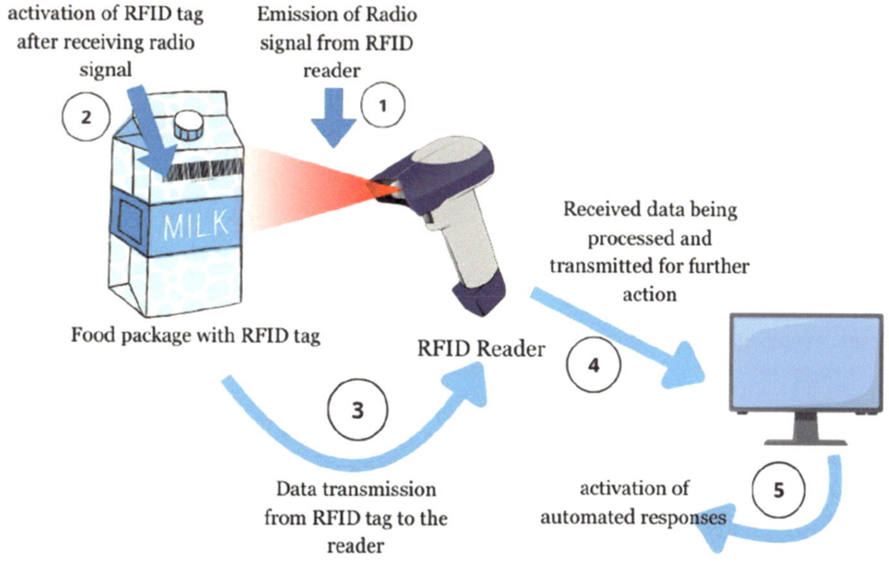

Figure 7.
Schematic diagram illustrating the working of RFID (adapted from the study by Azeredo and Correa [44]).

4. Consumer expectations and market drivers

Modern consumers demand packaging that is safe, sustainable, and informative. Smart additives play a key role in meeting these expectations [4].

4.1 Clean-label movement

Consumers prefer packaging made from natural, recognizable ingredients. This preference has led to a spike in biodegradable and edible packaging materials like chitosan [47], which offers antimicrobial properties and can be enhanced with essential oils or polyphenols [48, 49]. Numerous studies have demonstrated the effectiveness of such composite films in extending the shelf life of fresh and processed foods, including meats, seafood, and produce, without the use of synthetic preservatives [50]. These systems reduce reliance on synthetic preservatives while maintaining safety and shelf life.

4.2 Transparency and trust

Smart packaging builds consumer trust by offering visual or digital confirmation of product integrity. TTIs, freshness indicators, and traceability labels reassure buyers by offering visual or digital confirmation that a product has not been exposed to unsafe conditions during storage or transport [51, 52]. This aligns with global food safety regulations and traceability initiatives.

4.3 Convenience

Modern lifestyles demand food that is not only safe but also easy to use and assess. Intelligent packaging that offers visual freshness cues or mobile-based scanning options (via NFC/RFID) adds to the convenience factor. In this context, intelligent packaging systems that offer enhanced convenience through freshness indicators, quick-response (QR) codes, and mobile-based scanning technologies like NFC and RFID are gaining widespread acceptance [53, 54]. This supports better meal planning and reduces wastage.

4.4 Sustainability

Eco-conscious consumers favor packaging that is biodegradable, compostable, and non-toxic. Natural additives like turmeric, green tea extract, and essential oils are being incorporated into starch- or cellulose-based films to create functional, sustainable alternatives to plastic [55, 56]. Starch-based films containing turmeric extract have demonstrated effective antimicrobial activity while also being fully biodegradable, offering a promising alternative to petroleum-based plastics [56]. Brands that embrace such green technologies find greater acceptance among eco-conscious buyers, especially in premium retail segments.

5. Challenges and future outlook

Despite the promise of smart packaging technologies, several challenges need to be addressed to facilitate their widespread adoption in food systems.

5.1 High cost

Smart packaging systems, especially those incorporating sensors, indicators, or bioactive compounds, can increase packaging costs by 10%–20% compared to conventional materials. This is a significant barrier for low-margin commodities. However, the long-term benefits – reduced spoilage, extended shelf life, and enhanced product value – often justify the investment, particularly in premium supply chains. As manufacturing scales up and technologies mature, costs are expected to decline [57].

5.2 Consumer awareness

Many consumers remain unfamiliar with smart packaging concepts. Studies show that only 17% recognize the term "smart packaging." At the same time, consumers expressed gaining knowledge in the area of novel packaging and showed willingness to adopt the technology [1]. Misconceptions about chemical additives, misunderstanding of how indicators work, or fear of synthetic components can hinder widespread adoption. Beyond cost considerations, consumers expressed concerns about the potential risks associated with smart packaging materials, along with hesitation toward sudden changes in their purchasing habits. Therefore, educational efforts, transparent labeling, and consumer engagement campaigns are essential to build awareness and acceptance of these innovations, particularly in emerging markets [57].

5.3 Regulatory hurdles

Smart packaging additives must comply with food safety regulations, which vary across regions. Approvals from agencies like the Food Safety and Standards Authority of India (FSSAI), the Food and Drug Administration (FDA), or the European Food Safety Authority (EFSA) are required before commercial use. Additives involving nanomaterials, synthetic agents, or embedded electronics face stricter scrutiny due to migration risks. Manufacturers need to conduct rigorous safety testing to confirm that packaging components do not contaminate food and that the packaging remains safe under normal use. Harmonization of global standards and evidence-based safety assessment are critical for industry growth [58].

5.4 Waste management

Smart packaging often includes multilayer films, sensors, or electronic components that complicate recycling. These materials may not be biodegradable or compatible with existing waste systems. To address this, researchers are developing compostable smart films and circular packaging models. QR codes and freshness indicators can also guide consumers on proper disposal or reuse, reducing environmental burden [59].

6. Smart packaging additive systems in the post-harvest supply chain

Smart additives play a transformative role across the post-harvest supply chain, from farm to fork, enhancing quality, safety, and consumer engagement.

They include active and intelligent packaging that monitors the internal and external changes occurring in the product during storage and transport. These additives respond on the external surface by either electrical or optical means [60].

6.1 Harvest to sorting

Microbiological, physiological, and biochemical changes frequently cause the rapid decay of many high-moisture foods, particularly fresh herbs, fruits, and vegetables. They begin to undergo physiological processes like respiration, ripening, transpiration, senescence, and microbial destruction as soon as they are harvested. At this point, quality degradation can be considerably slowed down by the early application of additive-based treatment [61].

Precooling is done to remove the field heat from fresh fruits and vegetables, which slows down the biochemical and physiological activities of the produce [61]. For instance, ripening is postponed, and the microbial load is reduced by using pre-harvest or post-harvest sprays that contain natural antimicrobials or anti-ethylene compounds (such as potassium permanganate or 1-methylcyclopropene [1-MCP]). By ensuring the product enters the chain in ideal condition, these interventions lessen the strain on packaging additives and provide a cleaner, more stable platform for further processing [62].

6.2 Smart packaging (active and intelligent additives)

Food products are packaged with materials that have functional additives, such as coatings that release natural antimicrobials, sachets that absorb moisture or ethylene, or films built in with oxygen scavengers [63].

The purpose of active packaging is to extend the shelf life of the product and preserve the quality of the food in the pack by absorbing or releasing specific components from the packed food product or nearby foods [63].

The goal of antimicrobial food packaging materials is to improve shelf life while maintaining product quality and safety by extending the lag phase and inhibiting microbial growth [64].

TTIs, spoilage sensors, and freshness color strips are some of the intelligent components that are simultaneously placed outside the pack. These methods allow for external monitoring while preserving the quality of food. The packaging at this node serves as a dynamic system that not only works as a physical barrier but also actively interacts with the food and its surroundings in order to ensure safety and extend shelf life [65].

These systems provide external cues for monitoring while interacting with the environment and product to preserve quality. Compared to traditional methods like canning or MAP, smart packaging offers freshness, flexibility, and reduced reliance on chemical sterilants [65].

6.3 Cold chain transport and distribution

Temperature control is critical during the transportation and distribution stage, particularly for frozen or cold commodities. Food quality might be harmed, or spoilage can result from even minor differences. RFID and NFC tags embedded with sensors monitor temperature, humidity, and shock in real time. Data can be uploaded to cloud dashboards, enabling proactive quality control and immediate response to breaches in the cold chain [66].

6.4 Retail display

The retail display serves as a communication channel between the product and the customer by using intelligent additives that provide both protection and information. This encourages consumers to make educated decisions and minimizes wastage from the premature disposal of still-edible goods. TTIs and freshness indicators provide visual assurance of product integrity. Transparent films with antimicrobial coatings protect the product while showcasing its quality. These features reduce premature disposal and encourage informed consumer choices [67].

6.5 Consumer purchase and use

Consumers can scan NFC tags or QR codes to access detailed product information, such as harvest date, storage history, and safety status, through smartphones. This digital connection is especially crucial for consumers who are health-conscious, since it promotes a sense of transparency and control. Freshness indicators help reduce food wastage by showing whether a product is still safe to consume. Functional additives continue to protect the product during storage, ensuring safety and nutrition until consumption [68].

7. Additive benefits across the chain

7.1 Reduced spoilage

By using oxygen scavengers, antimicrobials, and ethylene absorbers, smart packaging actively delays spoilage, allowing food to remain fresh and safe for a longer duration. It also enhances the product's attributes, including its appearance, taste, flavor and aroma. This reduces food wastage during storage and transportation and helps to maintain a proper supply chain. Consequently, it helps consumers by providing food that maintains its freshness and quality until it reaches their table [69].

7.2 Extended shelf life

Smart packaging helps maintain food freshness and actively inhibits spoilage, thereby prolonging shelf life. Functional additives – including antimicrobials, ethylene absorbers, oxygen scavengers, and moisture regulators – when incorporated into packaging systems, actively inhibit microbial growth and reduce oxidation, thereby delaying quality degradation [13]. Controlled-atmosphere packaging environments can delay ripening, spoilage, and senescence by creating optimal storage conditions through the regulation of oxygen, carbon dioxide, and humidity levels. Smart packaging plays a prominent role in the preservation of fruits and vegetables. These techniques not only increase the shelf life of perishables but also enable longer marketing windows, which is advantageous for export and e-commerce platforms [70].

7.3 Traceability

Integration of traceability technologies in smart packaging offers a range of benefits across the supply chain, such as safety, minimization of wastage, and consumer trust. By using technologies like smart sensors, RFID, and QR codes, smart packaging can monitor real-time, end-to-end product tracking, enabling quick problem detection and direct consumer engagement. Smart sensors can measure and record environmental conditions, like temperature, humidity, or gas levels, inside the storage and transport units. RFID tags are sophisticated data information carriers, enabling quick tracing of a product's origin, processing, and movement without direct contact. QR codes offer detailed product information, such as batch numbers, certifications, and handling instructions, and allow both sellers and consumers to scan packages and access this information. Together, these technologies create a farm-to-fork chain of custody and add value in regulated or premium markets [71].

7.4 Clean-label compliance

The incorporation of natural, plant-based, or biodegradable smart packaging additives into products makes them safer and healthier and gives them fewer side effects compared to synthetic additives. Eco-conscious consumers claim that such additives are environmentally friendly, naturally obtained, and reduce synthetic or plastic waste. Health-aware consumers trust natural and plant-based additives more. Also, innovative and appealing packaging strongly impacts what consumers decide to buy. This helps brands align with clean-label trends, builds trust, and broadens appeal in today's conscious marketplace [72].

7.5 Consumer trust

Smart packaging, integrated with additives and technologies such as QR codes and sensors, enhances transparency and traceability from farm to table, ensuring product safety and authenticity, and thereby building consumer trust. Visible indicators, freshness cues, and digital transparency tools can identify spoilage, indicate freshness, trace the product's entire journey, and provide data on manufacturing processes and nutritional value, which foster a greater sense of confidence among buyers, reinforcing brand reliability and enhancing customer loyalty. However, introducing new technologies may require education and demonstration for consumers to understand smart packaging and ensure their willingness to use it [57].

8. Conclusion

As discussed throughout this chapter, smart packaging is an innovative technology impacting the quality, safety, integrity, and sustainability of food products. These systems significantly strengthen the post-harvest supply chain by enhancing transparency from harvest to consumption. Active packaging incorporates components such as antimicrobials, antioxidants, and gas scavengers to actively preserve food quality, extend shelf life, and reduce spoilage. It goes beyond passive protection by interacting with the food or its environment, improving safety and consumer confidence. Conversely, intelligent packaging employs sensors, indicators, or communication technologies such as TTIs, freshness indicators, RFID, and so on to track food quality, freshness, or storage conditions in real time. Unlike active packaging, it does not directly modify the food but provides valuable information to consumers and supply chain stakeholders to enhance safety and minimize wastage.

In the context of finite resources and a growing population, efficient utilization of available resources is imperative. These innovative technologies bridge the gap between ever-changing customer preferences and supply chain efficiency. Through their active and intelligent functionalities, smart packaging solutions extend shelf life, enhance product safety, and improve overall supply-chain efficiency. Although smart packaging additives help in minimizing wastage and ensuring transparency and traceability, there are some challenges. Cost, environmental impact, consumer trust, and regulatory approval are some of the main constraints limiting the widespread adoption of these technologies. Introducing new technologies may require educating consumers and providing demonstrations to foster understanding and encourage the adoption of smart packaging. Future research integrating science, industries, and consumer demands will be essential to fully utilize the potential of smart packaging.

Acknowledgments

The authors gratefully acknowledge the use of ChatGPT (OpenAI) for assistance with grammar checking, sentence restructuring, and enhancing the readability of the text.

Author details

Ravindra Naik[1*], Ramabhau Tumadu Patil[2], Aswathy Thankappan[1], Anjana Kallampilly[1] and Gowtham Manoharan[1]

1 ICAR – Central Institute of Agricultural Engineering, Regional Station, Coimbatore, India

2 ICAR-Central Institute of Post Harvest Engineering and Technology (CIPHET), Ludhiana, India

*Address all correspondence to: naikravindra@gmail.com

References

[1] Food and Agriculture Organization (FAO). The state of food and agriculture: Moving forward on food loss and waste reduction. FAO. 2019. DOI: 10.4060/CA6030EN

[2] Vasuki M, Kadirvel V, Pejavara Narayana G. Smart packaging—An overview of concepts and applications in various food industries. Food Bioengineering. 2023;**2**:e12038. DOI: 10.1002/fbe2.12038

[3] Brody AL, Bugusu B, Han JH, Sand CK, McHugh TH. Scientific status summary: Innovative food packaging solutions. Journal of Food Science. 2008;**73**(8):R107–R116. DOI: 10.1111/j.1750-3841.2008.00933.x

[4] Realini CE, Marcos B. Active and intelligent packaging systems for a modern society. Meat Science. 2014; **98**(3):404–419. DOI: 10.1016/j.meatsci.2014.06.031

[5] Müller P, Schmid M. Intelligent packaging in the food sector: A brief overview. Foods. 2019;**8**(1):16. DOI: 10.3390/foods8010016

[6] Asioli D, Aschemann-Witzel J, Caputo V, Vecchio R, Annunziata A, Næs T, Varela P. Making sense of the "clean label" trends: A review of consumer food choice behavior and discussion of industry implications. Food Research International. 2017; **99**(Pt 1):58–71. DOI: 10.1016/j.foodres.2017.07.022

[7] Sarma PP, Barman K, Baruah PK. Industrial barriers for the application of active and intelligent packaging. In Green Sustainable Process for Chemical and Environmental Engineering and Science. Elsevier; 2023. p. 71–96. DOI: 10.1016/B978-0-323-95644-4.00009-7

[8] Ramya S, Auxcilia J, Paital B, Sharmila DJS, Vethamoni PI, Venugopal S, et al Application of nanotechnology in fruit crops—from synthesis to sustainable packaging. Peer Journal. 2025;**13**:e19603

[9] Wołosiak-Hnat A, Zych K, Mężyńska M, Kifonidis A, Dajworski M, Lisiecki S, et al. LDPE/PET laminated films modified with $FeO(OH) \times H_2O$, Fe_2O_3, and ascorbic acid to develop oxygen scavenging system for food packaging. Packaging Technology and Science. 2019;**32**(9):457–469. DOI: 10.1002/pts.2449

[10] Gupta P. Role of oxygen absorbers in food as packaging material, their characterization and applications. Journal of Food Science and Technology. 2024;**61**(2):242–252. DOI: 10.1007/s13197-023-05681-8

[11] Cichello SA. Oxygen absorbers in food preservation: A review. Journal of Food Science and Technology. 2015;**52**(4):1889–1895. DOI: 10.1007/s13197-014-1265-2

[12] Duda-chodak A, Tarko T. Antimicrobial compounds in food packaging. International Journal of Molecular Sciences. 2023;**24**:2457. DOI: 10.3390/ijms24032457

[13] Kadirvel V, Palanisamy Y, Ganesan ND. Active packaging system—an overview of recent advances for enhanced food quality and safety. Packaging Technology and Science. 2025;**38**(2):145–162. DOI: 10.1002/pts.2863

[14] Geng C, Liu X, Ma J, Ban H, Bian H, Huang G. High strength, controlled release of curcumin-loaded ZIF-8/chitosan/zein film with excellent gas barrier and antibacterial activity for litchi preservation. Carbohydrate Polymers. 2023;**306**:120612. DOI: 10.1016/j.carbpol.2023.120612

[15] Drago E, Campardelli R, Pettinato M, Perego P. Innovations in smart packaging concepts for food: An extensive review. Foods. 2020;**9**(11):1628. DOI: 10.3390/foods9111628

[16] Gough R, Chevrier V, Tolbert M. Formation of liquid water at low temperatures via the deliquescence of calcium chloride: Implications for Antarctica and Mars. Planetary and Space Science. 2016;**131**:79–87. DOI: 10.1016/j.pss.2016.07.006

[17] Praxas. Calcium chloride vs silica Internet. Praxas; n d. Available from: praxas.com/en/calcium-chloride-vs-silica [Accessed: 2025-November-23]

[18] Flórez-Méndez J, López J. Food additives: Importance, classification, and adverse reactions in humans. Natural Additives in Foods. 2022;**17**:1–31. DOI: 10.1007/978-3-031-17346-2_1

[19] Sharma H, Mahajan G, Kaur M, Gupta R. Additives in dairy-based food. In InMicrobes for Natural Food Additives . Singapore: Springer Nature Singapore; 2023. p. 169–203. DOI: 10.1007/978-981-19-5711-6_8

[20] Singh S, Goswami MP, Vivekanand V. Effect of humectants on physico-chemical and sensory scores of hurdle technology-based shelf stable chicken pickle during storage period. DOI: 10.33545/26174693.2024.v8.i6Sh.1378

[21] Gliemmo MF, Campos CA, Gerschenson LN. Effect of sweet humectants on stability and antimicrobial action of sorbates. Journal of Food Science. 2004;**69**(2):fms39–44. DOI: 10.1111/j.1365-2621.2004.tb15513.x

[22] Sahu FM. Food additives: Making food taste better and safer. Beverages. 2016;**43**(6):20–28

[23] Lacroix C, Lachance O. Effect of various humectants and aw on proteolysis, yeast and mold growth and shelf-life during cold storage of yogurt. Canadian Institute of Food Science and Technology Journal. 1990;**23**(2-3):101–108. DOI: 10.1016/S0315-5463(90)70211-8

[24] Manalo M, Benedicto A. Effects of different humectants in the physico-chemical properties of ready-to-eat (RTE) tocino. InDLSU Research Congress 2014-March-6

[25] Aung SH, Nam KC. Impact of humectants on physicochemical and functional properties of jerky: A meta-analysis. Food Science of Animal Resources. 2024;**44**(2):464. DOI: 10.5851/kosfa.2024.e3

[26] Shankar S, Rhim JW. Polymer nanocomposites for food packaging applications. Functional and Physical Properties of Polymer Nanocomposites. 2016;29–55. DOI: 10.1002/9781118542316.ch3

[27] Gaikwad KK, Singh S, Negi YS. Ethylene scavengers for active packaging of fresh food produce. Environmental Chemistry Letters. 2020;**18**:269–284. DOI: 10.1007/s10311-019-00938-1

[28] Mariah MAA, Vonnie JM, Erna KH, Nur'Aqilah NM, Huda N, Abdul Wahab R, Rovina K. The Emergence and Impact of Ethylene Scavenger Techniques in Delaying the Ripening of Fruits and Vegetables. 2022;**12**:117. DOI: 10.3390/membranes12020117

[29] Chudasama MH. Ethylene scavengers in food packaging technology: A review. Food Chemistry. 2021;**337**:1–25. DOI: 10.13140/RG.2.2.17762.09923

[30] Chaemsanit S, Matan N. Activated carbon for food packaging application. Walailak Journal of Science and Technology (WJST). 2018;**15**(4):255–271. DOI: 10.48048/wjst.2018.4185

[31] Ayhan Z. Packaging and the shelf life of fruits and vegetables. DOI: 10.1016/B978-0-08-100596-5.22649-8

[32] Kamarudin SH, Rayung M, Abu F, Ahmad SB, Fadil F, Karim AA, Norizan MN, Sarifuddin N, Mat Desa MS, Mohd Basri MS, Samsudin H. A review on antimicrobial packaging from biodegradable polymer composites. Polymers. 2022;**14**(1):174. DOI: 10.3390/polym14010174

[33] Barbiroli A, Bonomi F, Capretti G, Iametti S, Manzoni M, Piergiovanni L, Rollini M. Antimicrobial activity of lysozyme and lactoferrin incorporated in cellulose-based food packaging. Food Control. 2012;**26**(2):387–392. DOI: 10.1016/j.foodcont.2012.01.046

[34] Dıblan S, Kaya S. Antimicrobials used in active packaging films. Food and Health. 2018;**4**(1):63–79. DOI: 10.3153/JFHS18007

[35] Soysal C, Bozkurt H, Dirican E, Guclu M, Bozhuyuk ED, Uslu AE, et al Effect of antimicrobial packaging on physicochemical and microbial quality of chicken drumsticks. Food Control. 2015;**54**:294–299

[36] Khaneghah AM, Hashemi SM, Limbo S. Antimicrobial agents and packaging systems in antimicrobial active food packaging: An overview of approaches and interactions. Food and Bioproducts Processing. 2018;**111**:1–9. DOI: 10.1016/j.fbp.2018.05.001

[37] Mkhari T, Adeyemi JO, Fawole OA. Recent advances in the fabrication of intelligent packaging for food preservation: A review. Processes. 2025;**13**(2):539. DOI: 10.3390/pr13020539

[38] Kalpana S, Priyadarshini SR, Maria Leena M, Moses JA, Anandharamakrishnan C. Intelligent packaging: Trends and applications in food systems. Trends in Food Science and Technology. 2019;**93**:145–157. DOI: 10.1016/j.tifs.2019.09.008

[39] Shetty JM. Time temperature indicators for monitoring environment parameters during transport and storage of perishables: A review. Environmental Conservation Journal. 2018;**19**(3):101–106. DOI: 10.36953/ECJ.2018.19313

[40] Wang S, Liu X, Yang M, Zhang Y, Xiang K, Tang R. Review of time temperature indicators as quality monitors in food packaging. Packaging Technology and Science. 2015;**28**:839–867. DOI: 10.1002/pts.2148

[41] Liu Y, Li L, Yu Z, Ye C, Pan L, Song Y. Principle, development and application of time–temperature indicators for packaging. Packaging

Technology and Science. 2023;**36**(10):833–853. DOI: 10.1002/pts.2761

[42] Skawińska E, Zalewski RI. Economic impact of temperature control during food transportation—A COVID-19 perspective. Foods. 2022;**11**(3):467. DOI: 10.3390/foods11030467

[43] Zhai X, Xue Y, Sun Y, Ma X, Ban W, Marappan G, et al. Colorimetric food freshness indicators for intelligent packaging: Progress, shortcomings, and promising solutions. Foods. 2025;**14**(16):2813. DOI: 10.3390/foods14162813

[44] Azeredo HMC, Correa DS. Smart choices: Mechanisms of intelligent food packaging. Current Research in Food Science. 2021;**4**:932–936. DOI: 10.1016/j.crfs.2021.11.016

[45] Lee SJ, Rahman ATM. Intelligent Packaging for Food Products. In Food Science and Technology, Innovations in Food Packaging. Han JH, editor. 2nd. Academic Press; 2014. p. 171–209. DOI: 10.1016/B978-0-12-394601-0.00008-4

[46] Magalhães AEV, Rossi AHG, Zattar IC, Marques MAM, Seleme R. Food traceability technologies and foodborne outbreak occurrences. British Food Journal. 2019;**121**:3362–3379. DOI: 10.1108/BFJ-02-2019-0143

[47] Jiang A, Patel R, Padhan B, Palimkar S, Galgali P, Adhikari A, Varga I, Patel M. Chitosan Based Biodegradable Composite for Antibacterial Food Packaging Application. Polymers. 2023;**15**(10):2235. DOI: 10.3390/polym15102235

[48] Cazón P, Velazquez G, Ramírez JA, Vázquez M. Polysaccharide-based films and coatings for food packaging: A review. Food Hydrocolloids. 2017;**68**:136–148. DOI: 10.1016/j.foodhyd.2016.09.009

[49] Cui H, Yuan L, Li W, Lin L. Edible film incorporated with chitosan and Artemisia annua oil nanoliposomes for inactivation of Escherichia coli O157:H7 on cherry tomato. International Journal of Food Science & Technology. 2017;**52**(3):687–698. DOI: 10.1111/ijfs.13322

[50] Sánchez-Ortega L, García-Almendárez BE, Santos-López EM, Amaro-Reyes A, Barboza-Corona JE, Regalado C. Antimicrobial edible films and coatings for meat and meat products preservation. The Scientific World Journal. 2014;**2014**:248935. DOI: 10.1155/2014/248935

[51] Yam KL, Takhistov PT, Miltz J. Intelligent packaging: Concepts and applications. Journal of Food Science. 2005;**70**(1):R1–10. DOI: 10.1111/j.1365-2621.2005.tb09052.x

[52] Anastasiadis F, Apostolidou I, Michailidis A. Food traceability: A consumer-centric supply chain approach on sustainable tomato. Foods. 2021;**10**(3):543. DOI: 10.3390/foods10030543

[53] Palanisamy Y, Kadirvel V, Ganesan ND. Recent technological advances in food packaging: Sensors, automation, and application. Sustainable Food Technology. 2025. DOI: 10.1039/D4FB00296B

[54] Kumari A, Gaikwad KK. Data carriers for real-time tracking and monitoring in smart, intelligent packaging applications: A technological review. Next Materials. 2025;**8**:100591. DOI: 10.1016/j.nxmate.2025.100591

[55] Shahid S, Leghari A, Farid MS, Saeed M, Anwar S, Anjum R, et al. Role

of active food packaging developed from microencapsulated bioactive ingredients in quality and shelf life enhancement: A review. Journal of American Science. 2021;**17**:12–28. DOI: 10.7537/marsjas170221.0

[56] Baysal G, Doğan F. Investigation and preparation of biodegradable starch-based nanofilms for potential use of curcumin and garlic in food packaging applications. Journal of Biomaterials Science, Polymer Edition. 2020;**31**(9):1127–1143. DOI: 10.1080/09205063.2020.1743947

[57] Nasution H, Harahap H, Julianti E, Safitri A, Jaafar M. Smart packaging based on polylactic acid: Effects of antibacterial and antioxidant agents from natural extracts on physical–mechanical properties, colony reduction, perishable food shelf life, and future prospective. Polymers. 2023;**15**(20):4103. DOI: 10.3390/polym15204103

[58] Agritech Digest. Challenges in implementing smart packaging for agri-products [Internet]. Available from: agritechdigest.com/challenges-in-implementing-smart-packaging-for-agri-products/ [Accessed: 2025-November-23]

[59] BL.INK. How smart packaging enhances sustainability and consumer engagement [Internet]. Available from: www.bl.ink/blog/how-smart-packaging-enhances-sustainability-and-consumer-engagement/ [Accessed: 2025-November-23]

[60] Joshi U, Bisht TS, Mamgain LR. Smart packaging: Modern way for reducing post-harvest losses of horticultural produce. International Journal of Agricultural Sciences. 2021;**17**(AAEBSSD):331–339. DOI: 10.15740/HAS/IJAS/17-AAEBSSD/331-339

[61] Hossain MA, Karim MM, Juthee SA. Postharvest physiological and biochemical alterations in fruits: A review. Fundamental and Applied Agriculture. 2020;**5**(4):453–469. DOI: 10.5455/faa.22077

[62] Fatima F, Basit A, Younas M, Shah ST, Sajid M, Aziz I. Trends in potassium permanganate (ethylene absorbent) management strategies: Towards mitigating postharvest losses and quality of mango (Mangifera indica L) fruit. Food and Bioprocess Technology. 2023;**16**(10):2172–2183. DOI: 10.1007/s11947-023-03047-8

[63] Duan Y, Wang GB, Fawole OA, Verboven P, Zhang XR, Wu D. Postharvest precooling of fruit and vegetables: A review. Trends in Food Science and Technology. 2020;**100**:278–291. DOI: 10.1016/j.tifs.2020.04.027

[64] Han JH. Antimicrobial Food Packaging. 2003. 50–70 p. DOI: 10.1016/C2013-0-18941-6

[65] Farber JM. Microbiological aspects of modified-atmosphere packaging technology—a review. Journal of Food Protection. 1991;**54**(1):58–70. DOI: 10.4315/0362-028X-54.1.58

[66] Pajic V, Andrejic M, Chatterjee P. Enhancing cold chain logistics: A framework for advanced temperature monitoring in transportation and storage. Mechatronics and Intelligent Transportation Systems. 2024;**3**(1):16–30. DOI: 10.56578/mits030102

[67] Lydekaityte J, Tambo T. Smart packaging: Definitions, models and packaging as an intermediator between digital and physical product management. International Review of Retail, Distribution and Consumer

Research. 2020;**30**(4):377–410. DOI: 10.1080/09593969.2020.1724555

[68] Young E, Mirosa M, Bremer P. A systematic review of consumer perceptions of smart packaging technologies for food. Frontiers in Sustainable Food Systems. 2020;**4**:63. DOI: 10.3389/fsufs.2020.00063

[69] Chen S, Brahma S, Mackay J, Cao C, Aliakbarian B. The role of smart packaging system in food supply chain. Journal of Food Science. 2020;**85** (3):517–525. DOI: 10.1111/1750-3841.15046

[70] Du L, Huang X, Li Z, Qin Z, Zhang N, Zhai X, et al. Application of smart packaging in fruit and vegetable preservation: A review. Foods. 2025;**14** (3):447. DOI: 10.3390/foods14030447

[71] Daoud MK, Trigui IT. Smart packaging: Consumer's perception and diagnostic of traceability information. In: International Conference on Digital Economy. Cham: Springer International Publishing; 2019. p. 352–370. DOI: 10.1007/978-3-030-30874-2_28

[72] Complete Packaging Group. The future of sustainable packaging – Innovations and trends to watch [Internet]. Available from: completepackaginggroup.com/uncategorized/the-future-of-sustainable-packaging-innovations-and-trends-to-watch/ [Accessed: 2025-November-23]